Fossil Identification Field Guide

Patrick Nurre

Fossil Identification Field Guide

By Patrick Nurre

Fossil Identification Field Guide
Published by Northwest Treasures
Bothell, Washington
425-488-6848
NorthwestRockAndFossil.com
northwestexpedition@msn.com
Copyright 2017 by Patrick Nurre.
All rights reserved.

Scripture quotations taken from the New American Standard Bible®
Copyright © 1960, 1962, 1963, 1968, 1971, 1972, 1973,
1975, 1977, 1995 by The Lockman Foundation
Used by permission. (www.Lockman.org)
Title page photo: Dinosaur vertebra, photo by Patrick Nurre.

Contents

Why We Should Study Fossils, and Tips for Collecting

Kids love fossils! It's that simple. I started collecting fossils when I was in second grade and have never lost the fascination with them since. Fossils have left evidence of a once mighty catastrophe that destroyed the surface of the earth about 4,500 years ago. And that is why it is important to collect and study the fossils!

Why collect and study fossils?

When you get right down to it, there are only two reasons to study the fossils:

1. **You believe that mankind came from a previous ancestor** who came from a previous ancestor and so on back to someone or something that originally looked nothing like me. This idea is known as *evolutionism*. (This is called a *secular* view.) In other words, man ultimately came from the simplest of cells through a complicated process of chance and change over hundreds of millions of years. All living things are related to one another and share a common ancestor. Secular scientists believing this to be true, diligently search for and collect fossils to bridge the gaps among the various living things to write the story of the evolution of all living things. The motivation of some secular geologists is also to escape the idea of being accountable to a God who has very clear moral values. Man naturally wants to be free from any kind of outside interference or accountability for the decisions he makes. Or...

4

2. **You believe all living things were originally created** as separate *kinds* of living things with the possibility of limited variation within their *kind*. This would allow them to adapt to changing environments but with strict boundaries. Kinds of living creatures only share relationship within their kinds. But a man is always and will always be a man and a bird is a bird and will always be a bird, no matter how many variations one finds of the same man or bird. In addition, most of the fossils we find in the rocks represent rapid burial and distribution by a global flood as described in Genesis. This idea is known as *creationism/catastrophism* and this is what Genesis teaches.

So, the motivation for collecting fossils is vastly different between the two ideas. The secularist collects fossils to reconstruct his evolutionary genealogy. The secularist believes the global Flood of Genesis to be a myth and not relevant to the study of fossils. The creationist/catastrophist collects fossils to understand the effects of the global Flood of Genesis and what life might have been like before the Flood. In other words, the fossil record is a record of mass extinction. The creationist/catastrophist believes the Genesis Flood was an actual historical event that came about because of the sin of man. **Fossils remind us that God will hold man accountable for his behavior.**

Tips for collecting fossils
Through the years of collecting fossils, I have learned a few simple lessons about collecting fossils:

1. You have to keep your eyes open and to the ground, always scanning the surface for telltale signs of fossils.
2. Collect everything, even if it looks insignificant.
3. Label your finds as to where you found them.

4. After you get them home, learn as much as you can about them. As fossils are much more fragile than rocks, you will need to be careful how you pack and store them. Use plenty of cushioning.

5. When you find a fossil, don't be preoccupied with trying to figure out what it is. Record the place of your discovery first. You can always study your fossil and identify it later. But you cannot always remember where you found it.

6. Often times you may think that you have found a dinosaur bone, but you are not sure if you have. I use this simple trick to help me decide if it is: I lick my finger and press it on the fossil. If it sticks slightly, it is likely a dinosaur bone.

7. Look for shapes and patterns in the rock. Do you see spirals or coils?

8. Great places to look for fossils include: river banks, gravel bars and pits, quarries, road cutouts, along beaches, and any kind of rock exposure, especially sedimentary. Remember that fossils can only be collected on private land, with permission.

Looking in a quarry or gravel pit

Along creek beds or along sandstone hills

Finds come in all sizes – a pachycephalosaurus bone and a fossil tooth

9. When consulting advanced fossil identification books, while they are often beautifully illustrated books and extremely helpful, they are written from an evolutionary point of view. Remember and practice the Basic Biblical Framework introduced in section V.

10. A good resource for locating fossil collecting sites is the **Roadside Geology** series of books. There is one for almost every state. They are evolutionary in their perspective, but do provide maps of outcrops of rocks and fossils along US highways. I take them with me wherever I go.

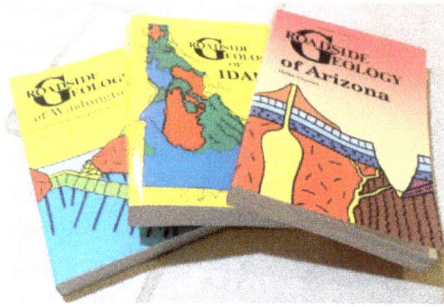

This book is a guide to help people of all ages identify what can be readily found in different localities, some of which are in their own backyard. I certainly found that out when I was a youngster walking home from school one day. There in the alley way lined with gravel, close to where I grew up, was a dinosaur bone, a big 50-pound dinosaur bone – my first big discovery! And ever since then I always have my eyes open looking in what might seem like the strangest of places. In this book I have recorded the fossils that I think kids will most likely find and a brief guide on how to identify them.

My goal is to help kids enjoy collecting fossils and to learn as much as they can about how they relate to the Genesis Flood as recorded in the Scriptures. The main thing to get out of this book, in addition to learning how to identify fossils, is to understand that fossils are remnants of the Genesis Flood.

I hope you enjoy this guide, and that your understanding of Genesis and the Flood is greatly enriched.

Happy fossil hunting,
Rock-man Pat

Please note: Many of the fossils in this field guide (from my personal collection) appear shiny. I have coated them with a preservative to protect them from disintegration while handling.

I. Learning the Fossil Terms

Often, fossil collecting and identification can be frustrating because geologists use technical words in the scientific languages of Greek and Latin. This alone is enough to discourage anyone. One thing I have never understood is that our culture now considers Latin and Greek to be dead languages, most likely because we don't use these languages in our everyday speech. But if you want to be a scientist, you must learn some Latin and Greek. It will not only help you understand scientific terms, but will even improve your English, too. So, it is well worth the effort to learn a few terms that will help you identify and enjoy collecting fossils.

Following is a short list of terms and words that will prove helpful in learning how to identify your fossils.

- **Biostratigraphy** – the branch of geology concerned with the separation and differentiation of rock units by means of the study of the fossils they contain. This is how secular geologists organize the rock layers. For example, all rock layers that have a particular kind of ammonite *(AM-mo-nite)* would all be the same age. But this ignores the Flood and assumes that the rock layers signify great ages of time. In contrast to this term, the Biblical geologist would refer to the study of the rock layers as Flood stratigraphy – the layers of rocks, and the Genesis Flood laid down the fossils in them, rapidly and consecutively. The prefix *bios-* is from the Greek, and has to do with *life*.

- **Calcite** – a common mineral consisting of the elements calcium, oxygen, and carbon (calcium carbonate) and is the most abundant mineral in sedimentary rocks.

- **Concretion** – from two Latin words, *con*, meaning *together*, and *crescere*, meaning *to grow*. Concretions are found in sedimentary strata. Although no one has witnessed concretions forming today, the fact that they are found in sedimentary layers (rocks laid down by water and mud), may hold a clue to their formation – the Flood of Genesis. Concretions are usually a sedimentary material such as clay, sandstone or limestone that has collected around a nucleus of some kind. Often the nuclei of these concretions consist of fossils of all types. Some people ask what the difference is between a *concretion* and a *nodule*. Although they may look similar, a nodule can be any material and is often volcanic in origin. A nodule is a lump, knot, or mass of aggregate mineral originally found in contrasting host stratum. An example of a nodule is a thunder egg found in hardened volcanic rhyolite ash or lava.
- **Coprolite** – meaning *dung stone*; the petrified/fossilized remains of *poop*!
- **Disarticulation (disarticulated)** – means a state of separation. The condition of being broken up and scattered. Dinosaur bones are most often found as disarticulated or separated bits and pieces of bone, usually in sandstone, siltstone, limestone, and other sedimentary rock. Disarticulation is a key to the Genesis Flood. The catastrophic nature of a global flood would have torn up most dinosaurs as they were initially overcome by the raging Genesis Flood, then buried under tons of sediments, and then torn up again as the Flood waters receded off of the face of the earth. This is why we rarely find complete dinosaur skeletons. Except for most marine/fresh water invertebrates and many vertebrates, most land vertebrate and plant fossils are preserved in the rocks as broken up pieces of bone. Why is this? No one knows for sure, but a

10

good guess is that most marine invertebrates and vertebrates that have been preserved as fossils were probably buried immediately with the breaking up of the fountains of the great deep. The last creatures to be buried in the Flood were most likely vertebrate land creatures that floated and then either gradually broke up as they began to decay or were eaten by surviving marine creatures. Later as the mountains rose and the valleys sank down according to Psalm 104:5-9, vertebrate land creatures that had been buried in the flood sediments were again broken up and transported across vast land masses. The word articulate is from the Latin *articulatus*, having to do with *dividing into joints*.

- **Extant** - means that something is still living. It is from the Latin *exstant*, meaning to be *visible or prominent or existing*.

- **Extinct – extinct** means that a creature is considered to no longer be in existence. The use of this term should be with caution, however. There are many examples of living things that have been declared to be extinct, only to show up somewhere in the world! Two familiar examples are the coelacanth *(SEE-la-canth)* fish, thought to have gone extinct 65 million years ago and then showed up alive off the coast of Madagascar in the 1930s and the Gingko plant which was once thought to be extinct, only to be discovered alive and well in China. It is from the Latin *exstinct* which means *extinguished*.

- **The Fossil Record** – the sum of the fossils that have been collected.

- **Fossilization** – the process of turning a once living thing into a fossil. The word fossil comes from a Latin word, *fossilis*, and means *obtained by digging*. This process is not fully understood today. The main reason is that not many dead things become fossils. They decay quickly, and that's the end of it! I don't know of

11

anyone who has ever witnessed something being naturally turned into a fossil. We have artificially turned things into fossils, but that is a long way from turning billions of things into fossils. What scientists do believe is that it takes **(1)** the right chemical environment, **(2)** preservation from oxygen and bacteria, and **(3)** rapid burial. Years ago it used to be said that for something to fossilize it took millions of years. That is no longer believed or taught. It takes the right *chemical environment* to form a fossil, not necessarily lots of time.

- **Glaciology** – the study of glaciers, their impacts and characteristics. The word glacier is derived from the French, *glace* which means *ice*. It is based on the Latin, *glacies*, which also means *ice*.

- **Ichno** *(ICK-no)* **fossil** – the word *ichno* is a Greek word, meaning *track*. It is not the fossilized animal or plant itself, but the evidence that it existed. It is also referred to as a *trace fossil*. So the imprint of a claw track would be a trace or ichno fossil.

- **Impression** – prints or marks made when an organism's body has been compressed or flattened. These can include tracks, skin impressions, tooth marks, claw marks, and casts.

- **Index Fossil** – any fossil thought to have first occurred in a particular geologic period, lived for a while and then either went extinct or evolved into something else. It is used as a type of time marker. For instance, when geologists first found a particular fossil in the record, it supposedly indicated the age of not only the fossil, but the sediments in which it was found. The problem has been that as more and more fossils have been discovered since the idea of index fossils was first described, the range of most of these index fossils has either been increased or decreased calling into question the whole use of index fossils as time indicators. But they are still adhered to by secular geologists today. The word *index* is from the Latin *index, indic-*, which means *forefinger, informer, or sign.*

- **Law of Faunal Succession** – a principle added to geology in the 1830s that interpreted the fossils in the sedimentary rock strata as having succeeded each other over time vertically in a specific, reliable order that can be identified over wide horizontal distances. Again, this principle has been used to interpret an evolutionary order to the fossils. An equally valid interpretive principle could see these fossils as having been buried in life zones, not necessarily in time zones. The word *fauna* comes from the Latin names *Fauna*, a Roman goddess of earth and fertility, the Roman god *Faunus*, and the related forest spirits called *Fauns*.

- **Law of Superposition** – an early principle introduced by Nicolas Steno in 1669 which states that in any given undisturbed group of layers of rock, the ones on the bottom were laid down first. This principle has been used by modern geology to imply great ages of time in the deposition and layering of the rocks. But this is simply an old-earth interpretation. Steno actually believed that these layers showed deposition by the Genesis Flood. The rock layers only show order of deposition, not how old they are. They do not show age. The prefix *super-* is from the Latin *super-*, which means *above or beyond.*

- **Lithification** – the little understood process of sediments turning to stone. The word is derived from the Greek word *lithos*, meaning *rock.*

- **Matrix** – the rock that surrounds the fossil. This is an interesting word, in that it is from the Latin, *mater*, meaning *mother.*

- **Mineralization (the same as permineralization)** – the process of replacing any organism's original material with a mineral. The two most common minerals found in fossils are silica (quartz varieties) and calcium carbonate (calcite varieties).

- **Petrification** – similar to fossilization, but more specifically it is the process that turns something to stone. It comes from the Greek word for *stone or rock, petra* or *petros.*

- **Sedimentary rocks** – those rocks laid down by water and mud, and containing most of the world's fossils. The word is derived from the Latin, *sedere,* meaning *to settle or sit.* These rocks include:

a) **Arkose** – a coarse-grained sandstone.

b) **Chalk** – a type of limestone consisting of calcium carbonate and tiny fossil skeletons of diatoms.

c) **Chert** and **flint** – hard silicified rock containing microscopic skeletons of diatoms, radiolarian, and algae.

d) **Coal** – a biochemical sedimentary rock made from the remains of plants and animals. The word *coal* is from an Old English form that means *mineral of fossilized carbon. Coal* is related to the German *kohl* (charcoal) and Irish *gual* (charcoal), and is attributed to the Indo-European "root" *g(e)ulo* (a glowing coal).

e) **Coquina** *(ko-KEE-na)* – a Spanish word meaning *shell.* It is a mass of limestone, disarticulated fossil shells, and bones.

f) **Limestone** – limy mud turned to rock.

g) **Mudstone** – another term for shale.

h) **Oil shale** – shale that is composed mostly of carbon.

i) **Sandstone** – tiny quartz crystals and iron held together by quartz, calcite and or iron.

j) **Shale** – very fine clay particles turned to stone.

k) **Siltstone** – tiny particles of sand or clay loosely held together by clay. The particles are much finer than sandstone, but larger than those that make up shale.

- **Strata** – a group of rock layers. It is derived from the Lain word, *stratum*, meaning *something that has been laid down.*
- **Stratigraphy** – the study of rock layers and what they mean. Again, it is derived from the Lain word, *stratum*, meaning *something that has been laid down.*
- **Stromatolites** *(stroh-MAT-o-lites)* – although not fully understood, these very interesting stone structures may actually be a type of fossil algae. The prefix *stroma-* is from the Latin, and means *layer or covering.*

There are two more terms that we need to look at, but they are so important that we are giving them their own chapter.

II. Adaptation and Kinds

Let's spend some time looking at **adaptation** and **kinds**. These two terms are crucial to understand if the fossil evidence is to be understood correctly.

Adaptation

Adaptation is the ability to thrive in one or more environments. The real question here is, did this ability evolve over millions of years in different creatures, or is there an ability to adapt to certain environments built into the genetic structure of living things, within certain limits, given to them at their initial creation in Genesis?

Do fossils tell us anything about adaptation? Yes, and no. Many plants and animals we find in the fossil record no longer exist. This is the sad consequence of the Genesis Flood. So the most that we can say about those plants and animals is that they were found in sedimentary layers of rock – rock laid down by water and mud. Since we cannot observe their life habits, we cannot say for sure much of anything about these plants and animals and their ability to adapt. Most of the massive fossilization we see in the fossil record was brought about by a massive global flood, not by a creature's inability to adapt to changing environments. We do observe extinction, but it was evidently because of the Flood, not because of an inability to adapt. Other plants and animals seem to have died out after the Flood. And this seems to be due to their inability to adapt to changing climate conditions resulting from the Flood.

The secular view and the creationist view of the fossils are two vastly different interpretations of what fossils tell us. Evolutionists view the rocks and the fossils in them as representing ancient life/time zones in which the plants and animals lived, died, and evolved over millions of years. As this has not been observed, it is simply an interpretation. On the other hand, the creationist views the fossils through the lens of the Genesis Flood. Creationists see fossils in sedimentary layers – period.

Does Genesis have anything to say about **adaptation**? Let's take the fifth Day of Creation as an example (Genesis 1:20-23).

> *Then God said, "Let the waters teem with swarms of living creatures, and let birds fly above the earth in the open expanse of the heavens." God created the great sea monsters and every living creature that moves, with which the waters swarmed after their kind, and every winged bird after its kind; and God saw that it was good. God blessed them, saying, "Be fruitful and multiply, and fill the waters in the seas, and let birds multiply on the earth." There was evening and there was morning, a fifth day.*

There are no minute subdivisions in the Genesis classification of living creatures. So, we have to tread very carefully. The evolutionists have made huge mistakes here and consequently have concluded some very silly things – like land animals evolving into whales, all without evidence to support that idea.

Without getting too technical, this is what we can conclude from this passage about Creation on Day Five: very simply, God created all the sea creatures, vertebrate and invertebrate, like whales, sharks, otters, seals, walruses, cephalopods *(SEFF-uh-lo-pods)*, gastropods *(GAS-tro-pods)*, corals; the great sea monsters could have included beasts like Ichthyosaurus *(ICK-thee-o-SAUR-us)*[1], Mosasaurus, megalodon, and the Plesiosaurs *(PLEE-see-oh-sore)*. He also created the flying creatures, feathered and non-feathered, like bats, birds, and Pteranodons *(ter-AN-o-don)*.

---

[1] Please note that some names in this text are capitalized, and some are not. This reflects the currently accepted nomenclature for naming living things.

In other words, creatures of the sea and air were created after their kinds with the ability to swim and fly. They were created with adaptation within their kinds. If you take some of the sea mammals for example and look at their individual traits, you get the following:

- **Whales** – warm-blooded, air-breathing mammals that cannot survive out of water.
- **Seals** and **walruses** – warm-blooded, air-breathing mammals that spend a lot of time on the land near the water, but are most active in the water.
- **Sea otters** – warm-blooded, air-breathing mammals that are equally at home in and out of water.

Evolutionists mistakenly believe that because these creatures live this way, they must have evolved the ability to adapt from land-living to sea-living creatures over millions of years. This has never been observed and raises more questions than answers. The only thing we can say is we only observe life habits as they exist today. You cannot draw the conclusion that they evolved into these habits. It may be true that some creatures have lost original genetic abilities, as flightless birds have, but originally, they had the built-in adaptation to fly.

Caudipteryx *(caw-DIP-ter-ix)* – an example of an extinct flightless bird

I point out all this because we must be careful not to lose sight of what fossils are – the remains of once-living creatures, some of which are now extinct but some creatures we have discovered as fossils are extant. One famous example of this is the coelacanth fish. Fossils may not tell us anything about adaptation, because we cannot observe those particular creatures and their living habits.

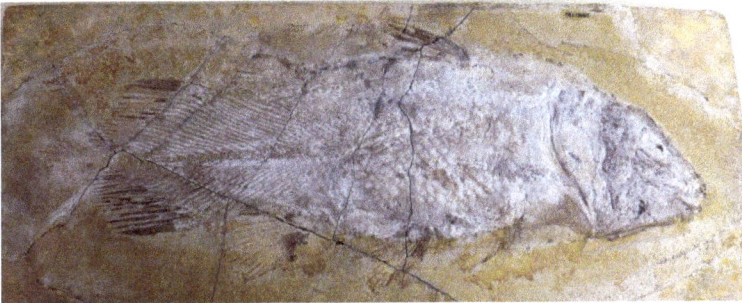

The fossil coelacanth *(SEE-luh-canth)*

Reconstruction of the living coelacanth

Kinds

The second term we need to examine is **kinds**. As the Scriptures do not give us much detail about how to subdivide kinds of living things, the only clue to defining kinds is found in the phrase, *after their kinds*. This would imply boundaries of some sort set up by God at Creation. The main clue could be *reproduction:* animals that can reproduce with one another. *Kinds* should not be taken to mean immutability (unable to change) as was mistakenly believed by Christians in the 1800s.

Within kinds there is ability for variation, but within its own kind. For example, there are many different varieties within the ammonite kind, which is also a kind within another kind, the cephalopods. But did ammonites have the ability to reproduce with baculites *(BACK-u-lite)* or scaphopods *(SCAF-o-pods)*, (also types of cephalopods)? We don't know, mainly because we cannot observe their reproductive behaviors.

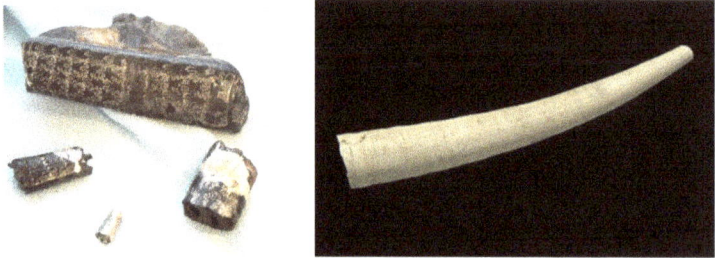

(Left) Baculite pieces – a type of cephalopod; (right) Scaphopod– a marine mollusk like the cephalopod

Many of the evolutionary ideas about plants and animals were popularized by Charles Darwin in the 1800's. He did not know, however, at the time he published his ideas, anything about genetics. Since his day, the science of genetics has grown significantly. With all the change that has taken place in individual animals and in populations of animals, evolution (change from one kind of creature into an entirely new and different creature) has never been documented. Change has taken place, but it has always been a downward trend – a loss of particular information that made a creature function in particular ways. A common example of this is the loss of the ability to fly in certain flightless birds.

So, some **kinds** of living things have gone extinct. Some **kinds** have exhibited adaptability to certain environmental changes.

But in all cases, there has been a limit as to the amount of change within a particular kind.

If we look at the trilobite *(TRI-lo-bite)* kind, as far as we know, it is extinct. Why did this kind of creature go extinct? We don't really know for sure. Flood geology would suggest that the environment in which it thrived was so changed by the Flood that it could not adapt. Secular geology would suggest that somewhere along the line the trilobite evolved into something else. But all we have to study are fossils of trilobites. And the only thing we observe is an amazing variety of trilobites, but no record that they evolved into something else or why they went extinct. The key word we need to keep in mind when we talk about *adaptation* and *change in kinds,* is **limits.**

The trilobite, now extinct as far as we know, shows an amazing number of varieties, but no evolution from one kind of creature to an entirely different one.

III. What is a Fossil?

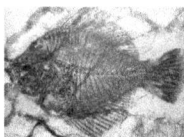

In the first chapter, we learned that the word *fossil* comes from a Latin word, *fossilis,* and means *obtained by digging.* At first, everything that was dug up was called a fossil. This included archaeological artifacts. Then the word came to be identified with the preserved remains or traces of animals, plants, and other organisms. Initially fossils were thought to represent plants and animals that had been destroyed in the great Flood of Noah's time, and this was the accepted explanation for hundreds of years.

Early in the 1800s however, as the book of Genesis was increasingly ridiculed and rejected as anything more than myth, fossils were seen as past extinct creatures that had either died out perhaps millions of years before or had changed into something new – evolution.

In the 1830s fossils were thought to be time markers. In other words, a fossil would show up in a series of rock layers and then disappear. These became known as Index Fossils, sorts of time markers that indicated the evolutionary time in which the creature lived. Early evolutionary geologists, therefore, thought that every rock layer that contained this fossil must be of the same age – everywhere in the world. But still the question persisted. *How old is the fossil?* By the time Charles Darwin developed his ideas about evolution, fossils were **thought** to be millions of years old.

It is important to note three facts about the age of fossils:

1. Fossils cannot be directly scientifically dated.
2. Fossil dating came about from an *idea* about evolution, not *actual scientific dates.*
3. Radiometric dating (the practice of dating rocks by radioactivity) did not exist until over 60 years after

Darwin, and it is based on certain accepted *assumptions*. (For a deeper understanding of radiometric dating, see Appendix A.)

So, what is a fossil? ***The most scientific thing we can say about fossils is that they are the preserved remains or traces of plants and animals that once lived in the past.*** And this may surprise you, but whether something is a fossil has nothing to do with age! That has been an idea implanted into our brains by evolutionists who believe that fossils represent past evolving relationships of living things preserved over millions of years.

Fossils can either be preserved in rock or can be found outside of rock. And with few exceptions, the only rocks that contain fossils are **sedimentary** rocks.

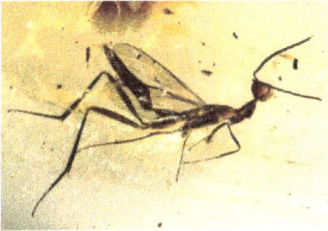

(Left) A fossil insect preserved in amber or tree sap; (right) A fossil fish preserved in limestone – a chemical sedimentary rock

Fossil wood preserved at Petrified Forest National Park (Arizona)

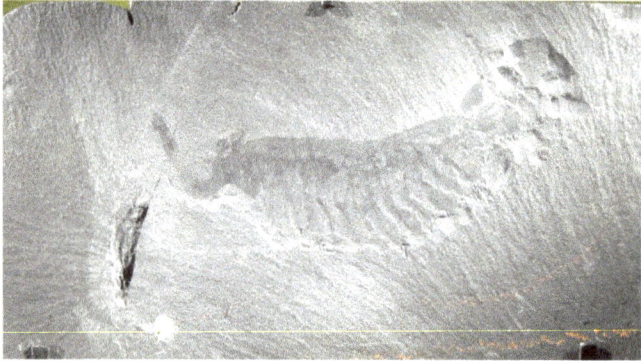

A fossil preserved in shale – a clastic sedimentary rock

Dinosaur bones preserved in sandstone – this is a clastic sedimentary rock; notice the scientists (in white) who are extracting the bones. These are huge dinosaur bones! (Colorado)

IV. Types of Fossils

There are a variety of fossil types. Some are completely overlooked because they don't stand out as fossils, but as strange markings. If you know the types of fossils, it may help you actually discover more fossils.

Types of Fossils

Permineralized (pronounced *per-MIN-er-uh-lized*) - Most fossils have been permineralized. That is, plant and animal cell structures have been replaced by some kind of mineral, like quartz or calcite. How did that happen? No one knows for sure because this process has not been observed on the scale we see in the fossil record. But we can make a good guess based on the Biblical account of the Genesis Flood.

a) Lots of mineral rich water would be needed for the once-living things to be soaked in. The Genesis Flood would definitely provide that! Where did the minerals come from? Most likely the minerals came from two sources. One source would have been the mineral rich pre-Flood earth. As the fountains of the great deep burst open, the water would have carried with it the minerals that had been stored up in the interior of the earth. Another source would have been the volcanic ash that would have been part of the breakup of the earth's crust. Volcanic ash contains lots of silica (quartz). And after this mixes with water, it produces mineral rich slurry that could have easily infiltrated the wood, cells, and bones of plants and animals caught in the Flood.

b) The right chemical environment would have been needed. A rapid burial of thick limy mud or other fine mud and sand would have preserved bones from decomposition.

The most abundant minerals involved in permineralization are quartz (found often as opal, agate, and jasper), calcite, and carbon. Permineralization means that you are not seeing the original bones or shells, but their bones or shells as their cells have been replaced by minerals.

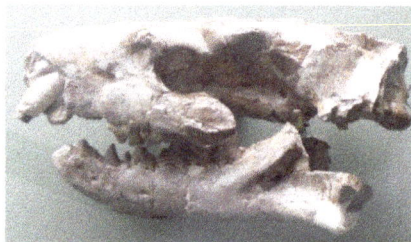

(Left) Clam petrified by calcite in limy mud - limestone made of the mineral calcite (Texas); (right) Skull replaced by silica-rich volcanic ash mixed with clay (Nebraska)

(Left) Turtle shell petrified in phosphorus-rich sediments (Florida); (right) Dinosaur bones petrified in silica-rich sandstone (South Dakota)

(Left) Insect preserved in very fine limestone –
limy mud and the mineral calcite (Green River Formation);
(right) The fossil wood was thoroughly saturated with silica-rich water.
Another name for this kind of petrified wood is agatized wood (Arizona).

(Left) These petrified gastropods have been replaced by quartz (agate). (Utah)
(Right) The ammonite has been replaced by the mineral pyrite. (Morocco)

These ammonites have been replaced by the mineral calcite.

Casts - Casts are not the remains of the animal or plant, but the traces of their existence. Casts show us that animals and plants once occupied the space in the mud, but then were removed either after leaving an imprint.

(Left) A cast of a starfish made in limy mud (Alabama) ; (right) A cast of a dinosaur track made in sandstone (Colorado)

Cast of a fern in mudstone (Illinois)

Molds – Molds are very similar to casts. But what is left are the permineralized remains of the animal or plant that had been surrounded by the mud or other sedimentary material.

Both a mold and cast of a trilobite made in mud (Bolivia)

Trace fossils – Trace fossils are not the actual remains of the plants of animals, but the traces of plants and animals. These fossils are called, *Ichno* fossils, meaning *tracks*.

A dinosaur track is an ichno fossil or trace fossil. Dinosaur tracks like this are also classified as casts. (La Sal Mountains, Utah)

These fossil worm tubes no longer contain the worms that once lived in them. They are therefore trace fossils or ichno fossils (Florida).

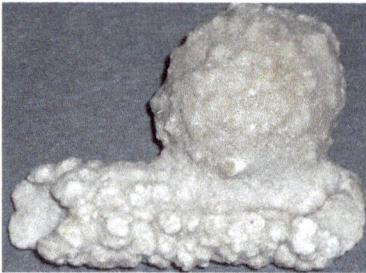

(Left) Fossil shrimp burrow (Florida): the shrimp no longer occupies the burrow, so it is a trace fossil or ichno fossil.
(Right) These ichno fossils are traces of worm burrows (Indiana).

If you will look closely you will notice tracks made by a trilobite. These are ichno fossils made in shale, a fine-grained sedimentary rock (Tennessee).

30

Soft-bodied fossils - Although rare, they do exist. Here is an example of one such fossil I found outside of Crawfordsville, Indiana.

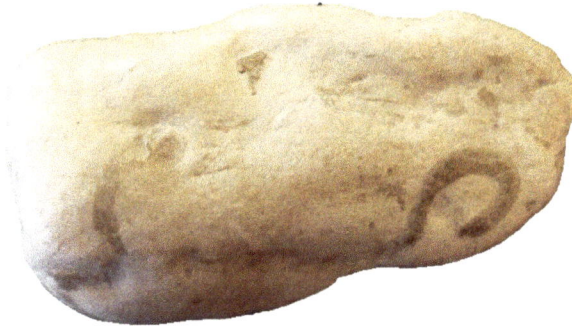

Fossil worms in limestone (Indiana)

Unaltered remains or the original remains – The most common type of these fossils is found in amber that is hardened tree resin.

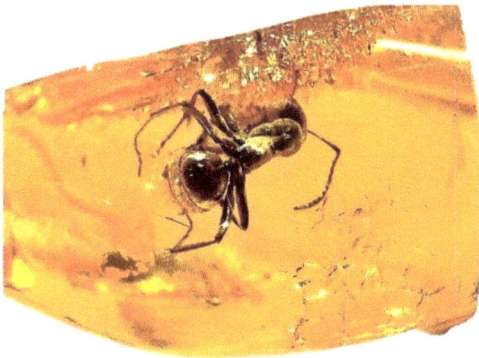

Carbonization or coalification – Carbonization is when only carbon remains in the original specimen.

(Left) Coal is the carbonized remains of plants and animals. (Right) Just how the immense coal beds of the world were formed is a secular geological mystery. But the Genesis Flood can explain it fairly easily through the wide distribution of plant mats sometimes buried between layers of other sedimentary rock. (Sydney Mines)

Unfossilized remains – These fossils are extremely rare, but have been showing up more and more recently. Fossils with soft tissue and blood vessels, and partially petrified remains of fish or bone have been reported in the scientific literature. One such example was reported in the news in 2007 by Dr. Mary Schweitzer, a paleontologist at North Carolina State University, who

> *is known for leading the groups that discovered the remains of blood cells in dinosaur fossils and later discovered soft tissue remains in the Tyrannosaurus rex specimen MOR 1125 [2,3], as well as evidence that the specimen was a pregnant female when she died.[4]*

Pseudofossils *(SOO-do-fos-ils)* – These are impressions or structures that look like fossils but in reality, are not. Some of the most mistaken of structures for fossils are *dendrites*. The

---

[2] Schweitzer, Mary H.; Wittmeyer, Jennifer L.; Horner, John R. (2007). "Soft tissue and cellular preservation in vertebrate skeletal elements from the Cretaceous to the present". Proc Biol Sci **274** (1607): 183–97. doi:10.1098/rspb.2006.3705. PMC 1685849. PMID 17148248.

[3] Hitt J (2005). "New discoveries hint there's a lot more in fossil bones than we thought". Discover. October. Archived from the original on February 22, 2007.

[4] "Geologists Find First Clue To Tyrannosaurus Rex Gender In Bone Tissue". Science Daily. 2005-06-03

word dendrite means *tree*. These impressions look like delicate branching ferns, but are not ferns at all. They are mostly produced by the mineral pyrolusite *(PY-ro-loo-site)*, made of the elements magnesium and oxygen (MnO_2 or magnesium dioxide). In its pure form pyrolusite looks like this:

Pyrolusite

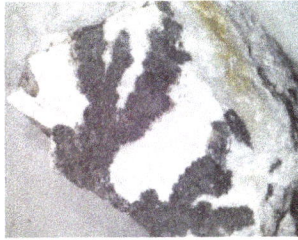

(Left) Pyrolusite forming as dendrites in dolomite (Utah); (right) A close up of the pyrolusite

Another common pseudo fossil is a concretion. The word *concretion* is derived from the Latin *con* meaning *together* and *crescere* meaning *to grow*. They can take on all kinds of shapes and sizes and may appear as fossils. But they are actually sedimentary structures usually formed around a smaller object. Now there is one thing to notice about concretions: they can

33

and often do contain fossils because fossils form the core of some concretions.

(Top left) Large concretion (about 12 inches across) on top of petrified wood (Montana); (top right and bottom) Concretions come in all shapes and sizes.

Sandstone concretions

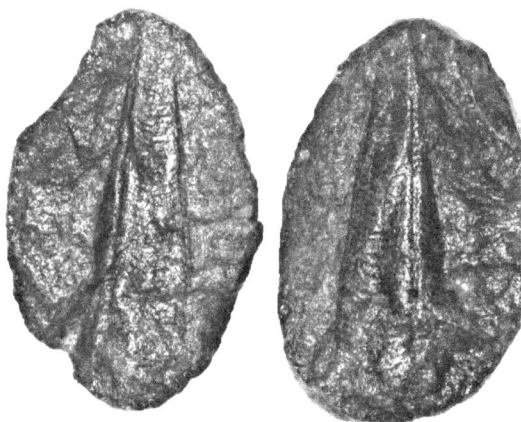

(Left) Partial fossil crab in a concretion (Washington); (right) Scaphopod (a cephalopod) and cast in a concretion (Bolivia)

Two other kinds of fossil types need to be discussed here because they are common but often overlooked in the field. They are **mud cracks** and **ripple marks**. Although technically not the remains of once-living things, they are preserved structural remains of past events and are therefore significant especially when talking about the Genesis Flood.

Mud cracks can be formed in two ways:

1. Desiccation or drying out: These mud cracks are quite common in arid areas of the country where

rain is not common. Soil dries out and cracks are formed.

2.   Compression: The type of mud crack seen in the picture below was formed from compression. Layers of mud were most likely repeatedly laid down one on top of another squeezing fresh mud into the layer below it forming what look like modern mud cracks, but in reality, are not.

Mud cracks – evidence of a lot of water and mud in the past (Montana)

Ripple marks are a bit of a mystery as to how they are formed. But in order for them to be preserved, they must be covered quickly so that they are not disturbed.

Fossil ripple marks (Montana)

V. A Biblical Model for Interpreting the Rock Layers Containing Fossils

 The Biblical model for interpreting the rock layers involves lots of moving water and sediments. Let's use Petrified Forest National Park as an example.

For years it was taught that the petrified logs in the Petrified Forest National Park were the remains of forests that grew in this area millions of years ago. Today it has been recognized that these petrified logs were actually moved into the area by massive flooding of some kind. They weren't buried here as ancient forests. How did they figure this out? Take a look at the pictures below.

In the distance, you can see petrified logs strewn about.

The picture in the foreground shows that the sandstone that once covered this area has been stripped by rushing water that left its flow marks and the presence of little bits and pieces of rounded stones. This kind of sedimentary activity is indicative of flood action.

Conglomerate sediments that indicate that they were laid down in rushing sediment-filled mud; notice the many rounded pebbles in the various petrified sediments above

This mess of petrified logs strewn about in the broken sandstone and clay/ash environment of Petrified Forest National Park indicates catastrophism, not slow and gradual evolution through millions of years. Notice the planation surface at the horizon. A planation surface is a broad, flat plain that indicates catastrophic erosion by rushing water. Secular geologists consider planation surfaces to be a major mystery, largely because they reject the idea of a global flood.

Rounded stones strewn about on the floor of Petrified Forest National Park are a sure indication of the tumbling action of floodwaters.

Petrified log surrounded by rounded, tumbled stones tell us that these logs did not grow here but were transported here by some kind of a massive flooding event

This petrified log has a covering of sediment filled with rounded, tumbled stones indicating that the log was buried in silica-rich, water-laid sediments. (The key is for perspective.)

Close-up picture of the petrified log covered with rounded, tumbled stones

41

As no one witnessed these various events that shaped the earth in the past and since there is no scientific way to settle the various interpretations of the fossils and rocks we have observed, we must construct a model for interpreting these phenomena. How are we going to do that? This can only be done if we choose some sort of philosophical/historical construct that we choose to interpret these things. There really are only two views that have been used to interpret the past history of the earth. And these have been at war with each other for the last 300 years. They can be summed up in the following two ways:

Uniformitarianism	Creation/Catastrophism
"Small and slow and long ago" [5]	*"Big and fast and in the recent past"* [5]
Geologic history that excludes the Genesis Flood and sees the various landforms of earth as a product of predominantly small and slow geologic processes acting over millions of years	Geologic history that includes the Genesis Flood and sees the various landforms of earth as a product of big and rapid geologic processes over a short period of time

If the Bible is actual history and more than nice pithy sayings and poetry, then constructing a Biblical framework should be possible and be able to supply an explanation for what we see around us. It does! Study the following diagram and use it as a template to interpret the various landforms you encounter on your family trips and vacations.

---

[5] I credit these two sayings to Dr. Gary Parker (ICR), as I have seen them only in his various books.

A Basic Biblical Framework

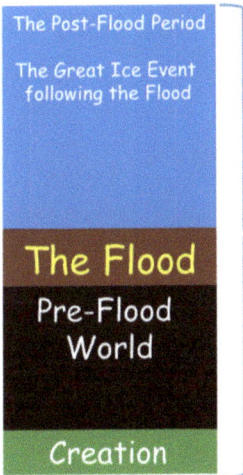

The Post-Flood Period

The Great Ice Event following the Flood

The Flood

Pre-Flood World

Creation

About 6,000 years

Because the framework has been revealed to us by God in the Scriptures, it is unchangeable. Our job as geologists is to look at any given landform and try to explain what we see in light of this framework. Secularists have their framework – uniformitarianism - and they attempt to interpret everything by it. And we have our framework. We must become familiar with how to use it to interpret the geology of the Earth. This will take practice, learning about rocks, and a keen understanding of the Scriptures.

Using the Basic Biblical Framework

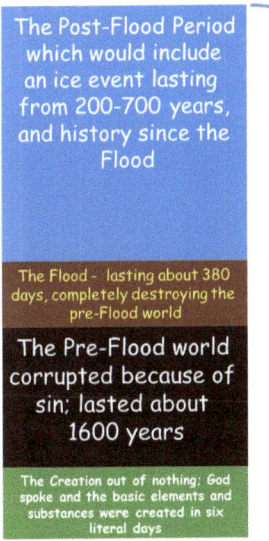

The Post-Flood Period which would include an ice event lasting from 200-700 years, and history since the Flood

The Flood - lasting about 380 days, completely destroying the pre-Flood world

The Pre-Flood world corrupted because of sin; lasted about 1600 years

The Creation out of nothing; God spoke and the basic elements and substances were created in six literal days

About 6,000 years

By taking a straight-forward understanding of the Book of Genesis and adding up the years given for the chronologies of the Patriarchs, one can establish a basic historical framework for interpreting earth history. It has always been amazing to me just how effective this simple tool has been at putting the pieces of geology together into a coherent whole. Even so-called mysteries of modern geology like planation surfaces and unconformities can easily be explained by following this simple Biblical Framework.

Geologists repeatedly tell us that fossils tell us a story of life struggle for existence. And because of this struggle, new life forms appeared over the millions of years of earth history. Geologists teach that fossil discoveries support this view. They tell us that that is why we find certain fossils only in certain layers and that these layers show an evolutionary progression from simple to complex. In light of our Biblical record of history, how do we handle this?

The **first** and probably the most important step is to realize that *all* past history which has not been directly observed is *interpreted* through a set of preconceived ideas. The observer may not even be aware of these preconceived ideas. But everyone has them. If a person is willing to honestly analyze his/her own thoughts, they will see just how much their own preconceived ideas influence how they see the evidence. In fact, preconceived ideas can influence a person to the point of leaving out certain things that might give a different picture. This has certainly been the case with modern geology. This is probably no more apparent than in the history of paleontology. There are many books on the market tracing the steps of the famous dinosaur bone hunters since the 1800s. This history is quite embarrassing and shows just how bias has influenced the way paleontologists have interpreted the fossil evidence only to change their views later as new interpretations are formulated.

A **second** step is to realize that the modern classification system of organizing fossils has become a tool for proving evolutionary relationships. We must get back to the book of Genesis and reexamine the word *kind*. Although we do not know exactly what a kind was in Genesis, we can see from the Biblical description that it involved boundaries and the ability to reproduce. This means that variation within a kind is not evolution, but an ability to adapt – within boundaries.

In the book, *Living Fossils*[6] by Dr. Carl Werner, he lists many examples of fossil plants and animals that were found in dinosaur rock layers that look strikingly similar to their modern counterparts! Thus, the fossil record does not show evolution, but variation. The other truth that the fossil record shows is extinction. There are many plants and animals that have disappeared. Was this due to the struggle to evolve? Or, did the Genesis Flood bring about mass extinction? Both of these views are interpretations and are influenced by what we believe about the world around us.

Some of the examples that Dr. Werner lists in his book include:

- **Invertebrates** including modern looking fossil insects, crustaceans, shellfish, starfish, crinoids, brittle stars, corals, sponges, earthworms, and marine worms – all appearing very similar to today's living invertebrate plants and animals, but found in dinosaur rock layers.
- **Vertebrates** including modern looking fish, sharks, rays, sturgeon, paddlefish, salmon, herring flounder, bowfin, hagfish, lampreys, frogs, salamanders, snakes, lizards, gliding lizards, box turtles, soft-shelled turtles, crocodiles, and alligators.
- **Birds** including parrots, owls, penguins, ducks, loons, albatross, cormorants, sandpipers, and avocets.
- **Mammals** including modern looking squirrels, possums, Tasmanian devils, hedgehogs, shrews, beavers, primates, and duck-billed platypus. According to evolutionary belief, mammals were supposed to have thrived *after* the dinosaurs became extinct. In fact paleontologists have found 432 mammal species in dinosaur rock layers! That's almost as many as the total number of dinosaur species paleontologists have described. A hundred complete mammal fossil skeletons have been reported as coming from dinosaur

---

[6] Dr. Carl Werner, *Living Fossils* (Masterbooks 2009).

rock layers. Dr. Werner states in his book that of the 60 museums he visited, he did not see a single complete mammal skeleton, found in dinosaur rock layers, displayed.

- **Plants** including modern looking flowering plants, ginkgoes, cone trees, moss, vascular mosses, cycads, ferns, sequoias, magnolias, dogwoods, poplars, redwoods, lily pads, and horsetails.

These fossil discoveries look almost indistinguishable from their contemporary living representatives. This fact is often masked, however, because paleontologists have given these fossil plants and animals different species and genera names which would certainly imply that there has been evolution taking place over millions of years. But remember, evolutionary biologists and zoologists determine the modern classification system. This might even be a case for circular reasoning!

Many of the fossil examples I will show in this book will look just like their modern counterparts. And so, I choose to interpret these fossils as evidence of the Genesis Flood – a recent historical, global geological event that drastically changed the face of the earth from what it had originally been like at Creation.

VI. Identifying the Invertebrate Sea Fossils, Part One

There are so many invertebrate sea fossils that we could cover, so we are going to take two chapters to cover them.

What is an invertebrate? An invertebrate is an animal without a backbone. There are several fossil animals that meet this criterion that are quite common in the fossil record:

- Trilobites
- Corals
- Crinoids
- Sea Urchins
- Bryozoa

One of the most amazing facts that paleontologists discovered as early as the 1840s was that what they called the lowest layer in the fossil-bearing strata contained billions of well-developed and well-diversified fossil animals. Life seems to have exploded on to the scene of life with no evolutionary history or ancestry. Most of the fossil animals mentioned above were part of what has been called, *The Cambrian Explosion*. This event remains today one of the most prolific unsolved mysteries of evolution. Darwin was extremely bothered by this phenomenon and considered it to be one of the main objections that could be made against his ideas of evolution by natural selection.

Even though paleontologists had considered these animals to be primitive in the evolutionary scheme of things, these animals were later shown to be highly developed creatures but showed no evolutionary history! Some of these animals have gone extinct but some still thrive in water environments today.

Where to find these fossil sea animals?

Since these were marine animals, you can find these creatures in almost any area that sedimentary rock exists, especially in limestone and in shale.

Limestone is generally a light-colored rock. But it can be darker in color too. Look for fossils. Also, if you have some muriatic acid, if after placing a drop on the rock it fizzes, you have found limestone.

(Left) Limestone with fossils; (right) Limestone without fossils

(Left) Limestone with fossils; (right) Limestone without fossils

Another form of limestone is called **coquina**. Coquina is a mass of cemented fossil shells and is quite common on the coasts of Florida and California.

Coquina – one containing fossil shells (Florida), the other a fossil vertebrate bone (Texas)

Shale is a fine-grained sedimentary rock usually exhibiting layers and can contain an abundance of fossils.

Samples of shale from Montana, California, and Nebraska

Shale formation (near Anaconda, Montana)

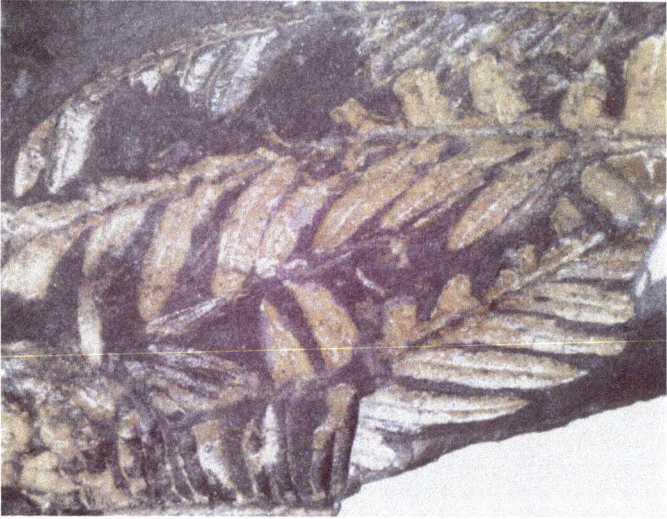

Fossil ferns in carbonized shale (the whole piece is seven inches long)
(Pennsylvania)

Sandstone can also contain fossils. Sandstone is a coarse-grained, sedimentary rock, made up primarily of tiny quartz crystals. Some fossils you may find in sandstone could include mud cracks and ichno fossils.

(Left) Sandstone formation (Utah); (Right) Ripples in sandstone (Utah)

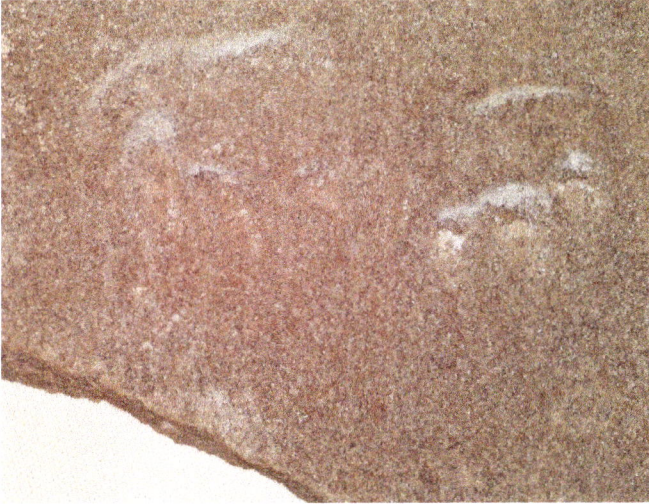

Tracks. about one inch wide, in sandstone, seen at upper left and right
(Arizona)

Invertebrate and vertebrate **sea animals** were created on Day
Five of Creation Week.

> *Then God said, "Let the waters teem with swarms of living
> creatures, and let birds fly above the earth in the open expanse
> of the heavens." And God created the great sea monsters and
> every living creature that moves, with which the waters swarmed
> after their kind, and every winged bird after its kind; and God
> saw that it was good. And God blessed them, saying, "Be
> fruitful and multiply, and fill the waters in the seas, and let
> birds multiply on the earth." There was evening and there was
> morning, a fifth day.*

Trilobites are numerous in the fossil record. The word means *three lobes* and that is primarily how they are identified. Their variety seems almost limitless, and yet they are all classified as trilobites. The basic identifying characteristic is the body style – three lobes, with a few exceptions. Take a look at some of the types of trilobites you might find.

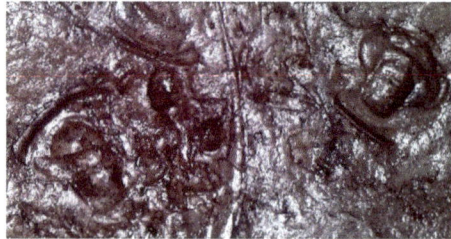

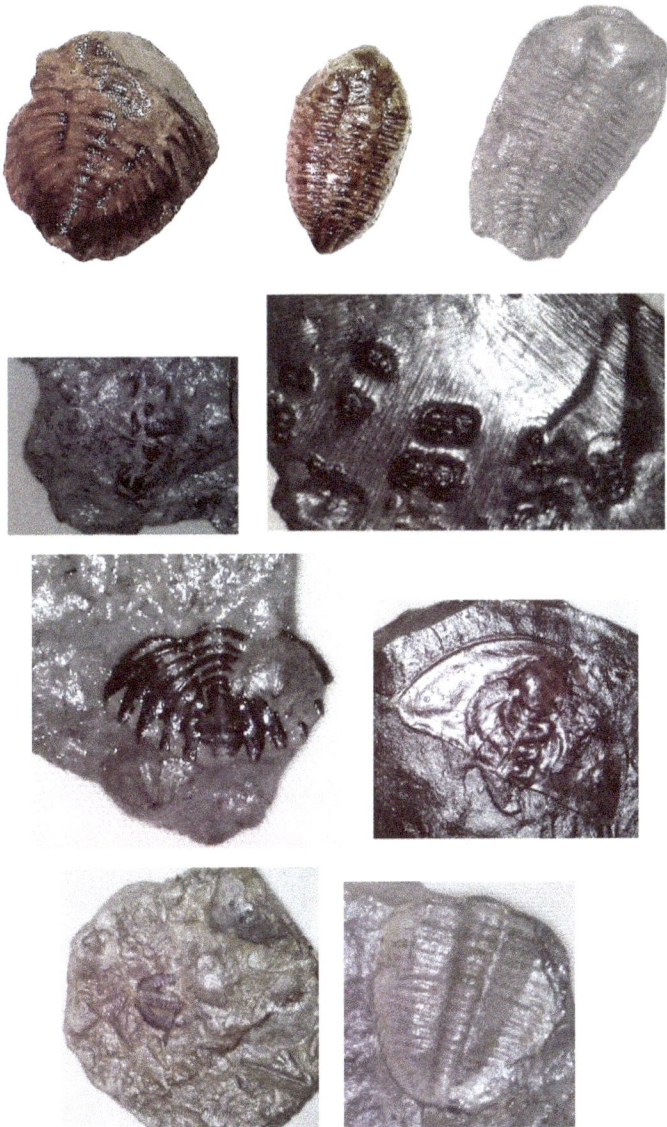

Trilobites

Corals are colony animals characterized by having polyps, which are essentially cylinders that have housed the internal organs of the individual animal. There are a huge variety of corals both **extinct** (not existing anymore) and **extant** (living today).

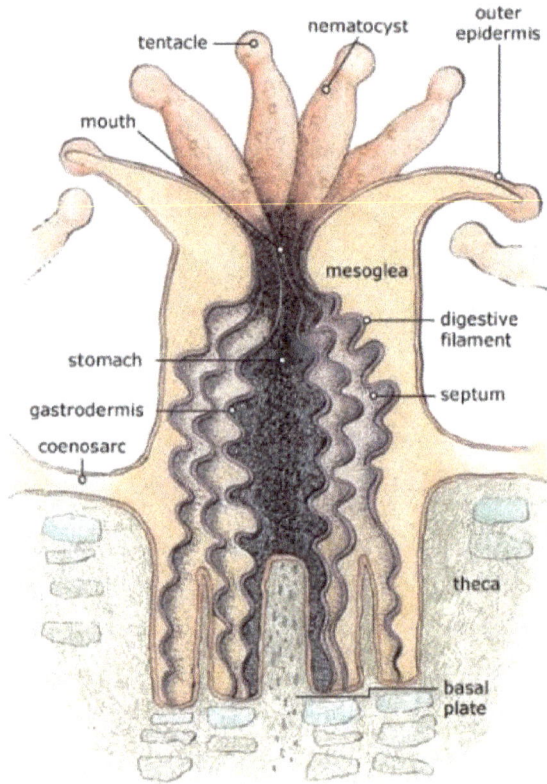

Anatomy of a live coral polyp

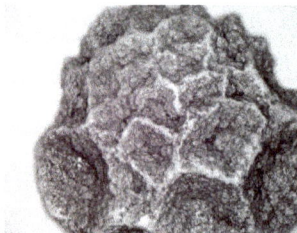

Fossil corals

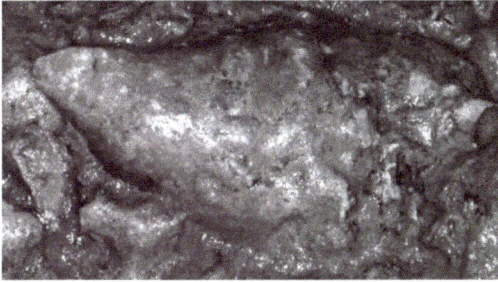

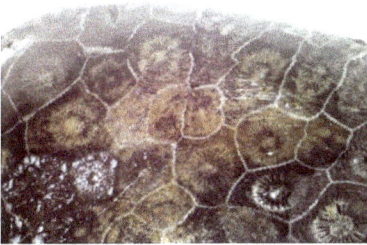

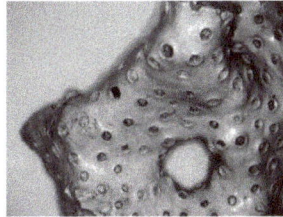

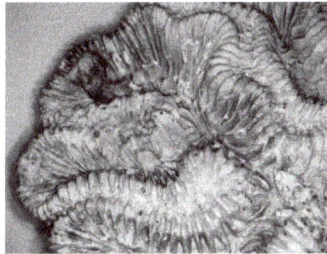

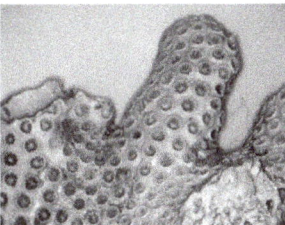

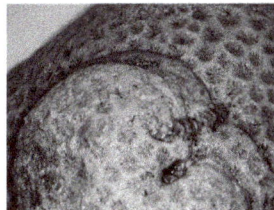

Fossil corals

Crinoid *(CRY-noid)*– comes from the Greek word, *krinon* (lily), and *eidos* (form). So, the word means *lily form*, which might imply that it is a plant. But that is a misnomer. Crinoids are animals and belong to the echinoderm family of which the starfish and sea urchins belong.

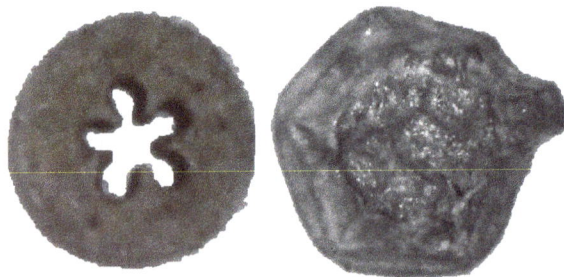

Each crinoid piece (about ¾ inch across) bears a five-pointed star shape. This is the identifying signature of the Echinoderms. Starfish, sea urchins and sand dollars all have it.

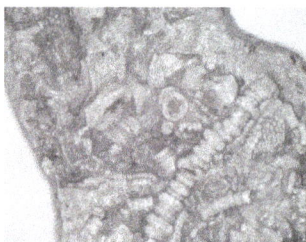

Most crinoid fossils are found broken up or disarticulated, a casualty of the Genesis Flood.

Crinoids

Crinoids

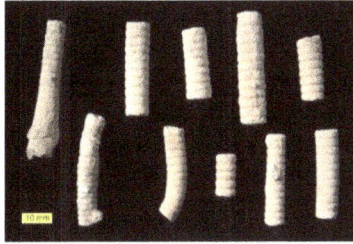

(Left) Crinoid Hold-fasts – a root-like structure for anchoring the crinoid stem; (right) Crinoid Stem Pieces

It appears that many crinoids went extinct during or shortly after the Flood. There are some extant kinds living today, however.

A Living Crinoid

Sea urchin – the word means *hedgehog* because the living sea urchins have a covering of spines that makes them look like hedgehogs. They are actually very complex organisms and continue to thrive today.

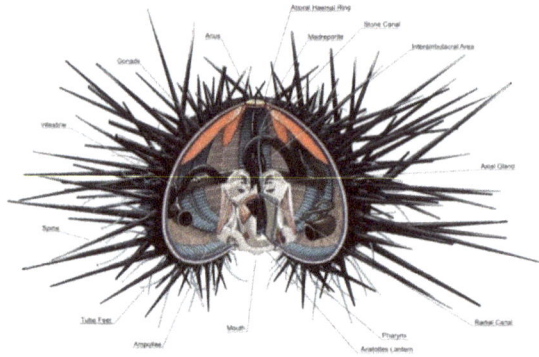

A living sea urchin

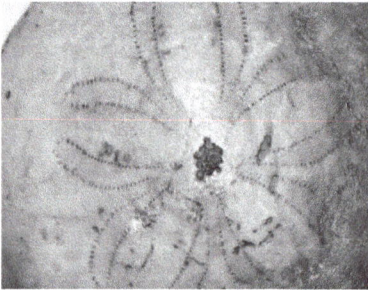

(Top left) Fossil sea urchin; (top right) fossil sea urchin spines, about one inch long; (bottom) fossil sand dollars (Florida)

Bryozoan *(BRY-uh-ZO-un)* – can be a colony animal similar to corals. In fact they are often mistaken for corals, but they don't form huge reefs as corals do. They are sometimes referred to as moss animals. They are filter feeders. Although many are extinct today, there are some extant varieties.

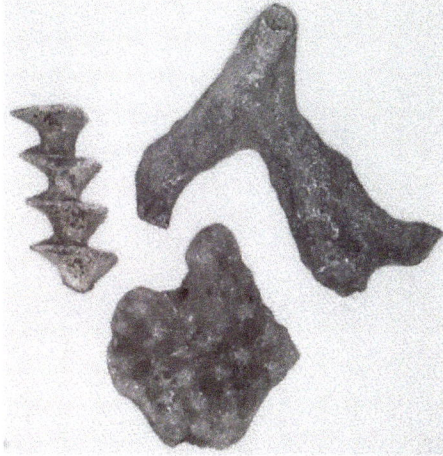

Fossil bryozoan varieties about ¾ inch long (Ohio)

A mortality plate five inches long (*graveyard fossils*) of bryozoan (Indiana)

59

VII. Identifying the Invertebrate Sea Fossils, Part Two

Let's continue on with looking at the invertebrate sea fossils. In this chapter, we will look at:

- Brachiopods
- Pelecypods or bivalves
- Gastropods
- Cephalopods
- Protozoa

Brachiopods *(BRACK-ee-oh-pods)* – the word brachiopod means *arm-footed*. They are also known as *lampshells* because of their uneven shell structure.

Live brachiopod, showing arm in white

As with many sea animals, some are extinct and some are extant. The picture below shows what we call a *graveyard fossil* or *mortality plate*, indicating millions of creatures that all suffered the same water catastrophic fate and were buried together. This is extremely common in the fossil record and these are the kinds of fossils that you will find the most of.

Mortality plate of brachiopods (Ohio)

Many people confuse brachiopods with clams and oysters. Brachiopods are their own kind of animal. There is a great way to tell the difference between the brachiopods and the clams. The answer lies in the arrangement of their shells. If you can find a whole clam and a whole brachiopod, the brachiopod's shells have what appears to be an *overbite*. Brachiopods are also called *lampshells* because they look like an old-fashioned lamp.

The following fossil brachiopods range in size from 1 to 1 ¼ inches long – notice the *overbite* and lamp shape. In these pictures, the overbite is apparent at the top of each picture.

Brachiopods

Pelecypods *(puh-LEE-see-pods)* **or bivalves** – comprise a class of marine and freshwater **mollusks** that have laterally compressed bodies enclosed by a shell consisting of two hinged parts. Mollusks in general have soft parts surrounded by shells made of calcium carbonate. Bivalves include **clams**, **oysters**, **cockles**, **mussels**, **scallops**, and numerous other families that live in saltwater, as well as a number of families that live in freshwater. The majority are filter feeders. Bivalves belong to a huge phylum called mollusks that include: pelecypods, cephalopods, gastropods, and scaphopods, among other lesser-known animals. Mollusks are thought to comprise around 23% of all known marine animals. That is a huge number. And this may explain why their fossils are so numerous in the fossil record.

Giant clam

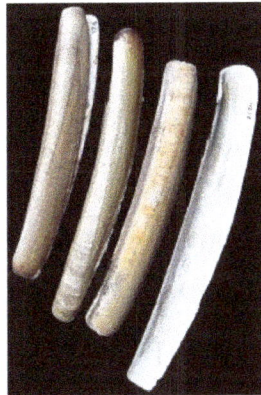

Razor clams

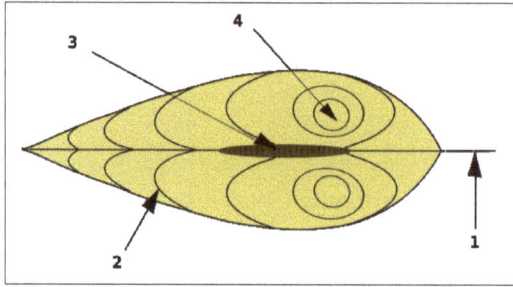

The main observable difference between a clam and brachiopod is that the shells of the clam are symmetrical.

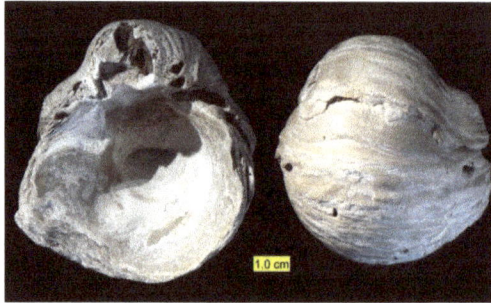

This type of fossil oyster is affectionately known as a *Devil's toenail* – a type of mollusk.

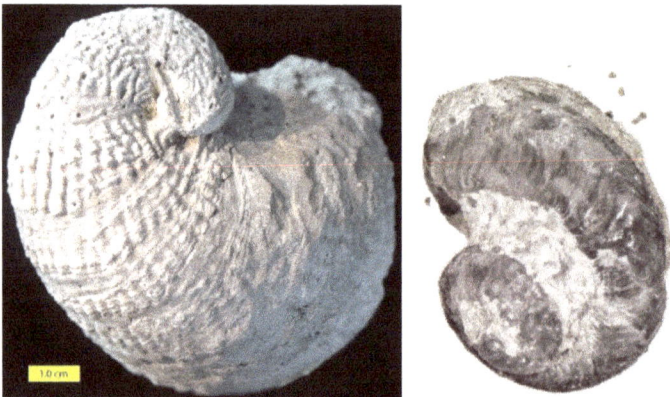

Fossil oyster shells

Fossil mussel (Oregon)

Gastropods – the word gastropod means *stomach footed*. It looks like it walks on its stomach. There is an enormous variety of gastropods both living and in the fossil record. The gastropod may look like a very simple creature, but it is not. It is complex having organs, eyes, a mouth, stomach, heart and very interesting and colorful shells. Gastropods today comprise around 80% of all mollusks.

Mortality plate about four inches across (*graveyard fossils*) of gastropods (Utah)

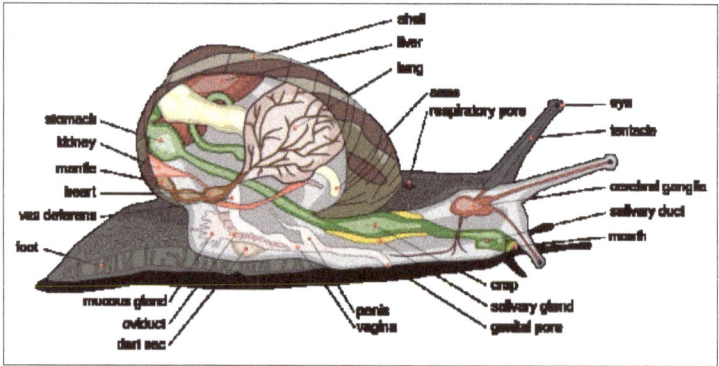

Anatomy of gastropod or snail

Fossil gastropod (turritella), a type of mollusk

Flattened snail fossil

Each of these fossil snails is about ¾ inch long (Texas).

Fossil gastropod or snail, about five inches long (Texas)

Various mollusk fossils

Cephalopods – the word cephalopod is a Greek word meaning, *head-footed*. It looks like it walks on its head with its long tentacles looking much like its legs. Cephalopods are mollusks. There fossil record is enormous with lots of variety. But apparently, the Flood was a major factor for much of their disappearance. Why? We don't really know. Perhaps the temperature of the oceans changed significantly after the Flood, causing an environment that just was not friendly to their survival. We can only admire the variety of the cephalopods from their fossils. The cephalopods of today include the nautilus, octopus, squid, and the cuttlefish.

The greatest variety in the fossil record is among the ammonites, a type of cephalopod. The name *ammonite* was actually coined by Pliny the Elder of Pompeii fame in the first century AD. He called fossils of these animals *ammonis cornua*, meaning *horns of Ammon* because the Egyptian god Ammon or Amun was typically depicted wearing ram's horns. The nautilus seems to be the only living representative of the ammonite.

A living nautilus – shelled cephalopod

Common living octopus

Common living squid

Common living cuttlefish

Take a look at these various ammonites.

Baculites – uncoiled cephalopods

Fossil straight cephalopods

Fossil graveyard of ammonites (cephalopods)

Because of the great amount of variation observed in the fossils among the cephalopods, geologists have concluded that they must have evolved the ability to adapt to different environments and therefore evolved this adaptation resulting in great variation. But that is a misinterpretation based on naturalistic thinking and a rejection of the Biblical historical record.

Another way to interpret this variation is that God created kinds with limited variation and with the ability to adapt within certain limits to various environments. And the fossil record could be interpreted in this way. What we don't see is the fossil evidence that would show one creature evolving via adaptation into something entirely new.

Crustaceans – form a very large group of arthropods, usually treated as a subphylum, which includes such familiar animals as crabs, lobsters, crayfish, shrimp, krill, and barnacles. They are very complex animals and show very little change from their fossil counterparts.

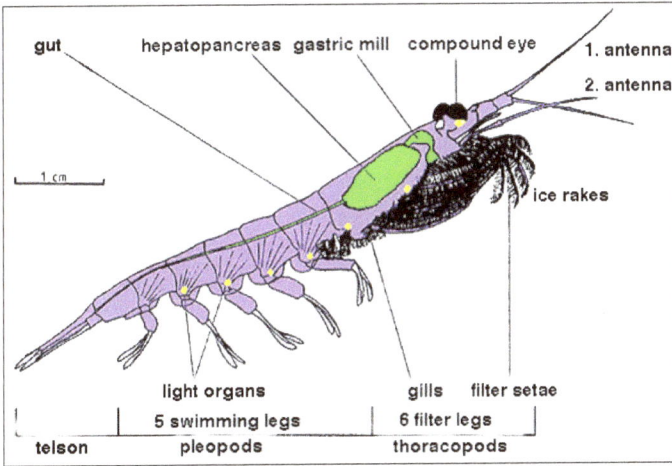

A typical crustacean structure

Fossil shrimp in limestone about two inches long (Green River Formation in Wyoming)

This fossil lobster was found in dinosaur rock layers! This means two important things. First it means that there has been no evolution in the lobster, as it looks exactly like its living counterpart. Second it means that it survived the supposed great dinosaur extinction! How can that be?

(Left) Crab fossil (Denmark); (right) Barnacle fossil (Maryland)

Protozoa – this word is no longer used for these particular animals. Why is that? The name in Greek means *first animal.* It was believed in the 1800s that these animals were primitive animals that evolved from something lower into the higher classes of multi-celled animals. Since then it has been discovered that these creatures were highly complex living creatures of their own kind. *Evolution evolved!* These living creatures show design! As fossils, they are often oval-shaped and look like little rice grains. This will better help you identify the fossil if you find something that looks like it.

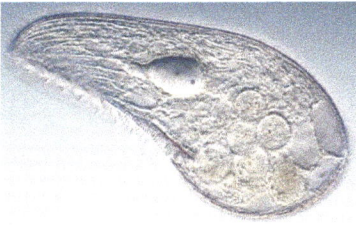

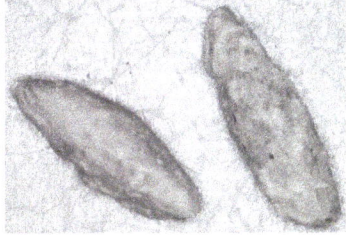

(Left) A living protozoan; (right) Fossil protozoa (about 1/8 inch long) (Texas)

Fossil graveyard of fossil protozoa in limestone, about two inches long (Texas)

VIII: Identifying
the Vertebrate Sea Fossils

 Vertebrate sea animals were created on Day Five of Creation Week. They include sea animals with backbones. Some of the more common fossils you will find among these creatures are:

- Shark teeth and vertebrae
- Megalodon teeth
- Mosasaur teeth and bones
- Fish, fish vertebrae, scales, and teeth
- Dugong bones (like the modern manatee)
- Whale bones and parts
- Ray teeth and stingers

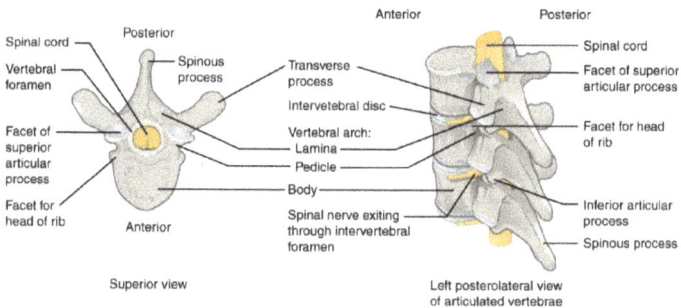

Vertebral Column shared by all vertebrate animals – a design of the Designer, with many variations

Next to the invertebrate sea creature fossils, the most abundant fossils seem to be those of the vertebrate sea creatures. There are lots of varieties of these fossils. The most prolific fossils are the shark teeth. Why? Sharks grow as many as 30,000 teeth in their lifetime. And fossil sharks are amazingly similar to their extant varieties. Shark teeth are found in a lot of places around the world. But the shark tooth capital of the world seems to be Florida. If you ever go to Florida, the inland rivers are loaded with fossil shark teeth. The oldest record of collecting shark teeth is from Pliny the Elder (AD 23-AD 79) who believed that these triangular objects fell from the sky during lunar eclipses.

Shark teeth (Israel)

The Megalodon *(MEG-uh-lo-don)* – the word means *big tooth* and it is appropriately named! Its large teeth are easy to purchase but depending on their size and condition can be very expensive ranging from a few dollars to upwards of $1000 each. The largest megalodon tooth found has been nine inches long.

Many think that megalodon was a giant shark ranging in size up to 50 feet long. Geologists think that because very few of its vertebrae or other hard parts have been found. This has led to the belief that megalodon was built mostly of cartilage. This is a good guess, but we don't know for sure.

Megalodon tooth compared to a couple of Great White shark teeth

Boxes of megalodon shark teeth for sale! These teeth are not rare.

Fossil megalodon teeth

The bull shark like its fossil representative can have as many as 50 rows of teeth, but most sharks have five rows with the average shark having about 15 rows of teeth in each jaw.

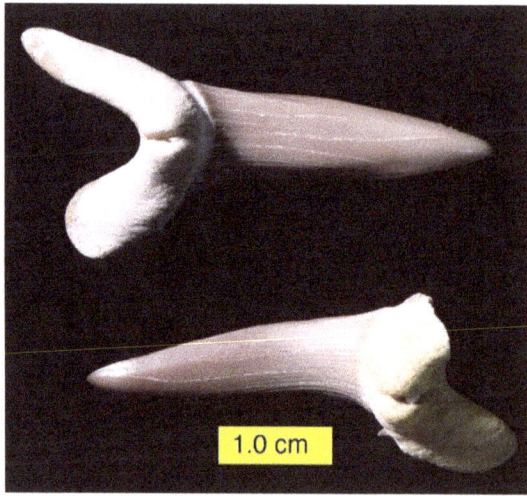

Shark teeth

Fossil shark teeth from Florida compared to the megalodon tooth in the upper left of the picture, which is two inches long

The Mosasaur *(MOE-suh-saur)* or Mosasaurus – The word means *lizard of the Meuse River* where it was first discovered. What was this strange creature? Some have called it a toothed whale. Some have called it an aquatic lizard. Whatever it was, it is apparently extinct now. But it was huge! It was as much as 50 feet long. Its fossils are abundant in Texas, Kansas and Morocco. The teeth are quite common and affordable as well as its bones and vertebrae. Most people think that the mosasaur went extinct 65 million years ago. But here is something interesting worth noting: Georg von Forstner, a German submarine Captain in World War I, reported seeing what he described like a 60-100-foot Mosasaur shooting out of the water. So, is this creature really extinct?

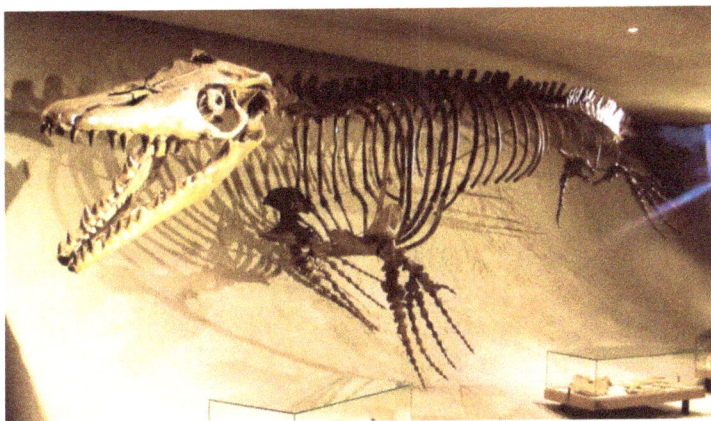

Mosasaur skeleton

Mosasaur tooth (Morocco)

Characteristic marine reptile bones, about two inches long (Texas)

Fossil gar scales, about 3/8 inches long (Florida)

Fossil fish showing scales and fins, exquisitely preserved

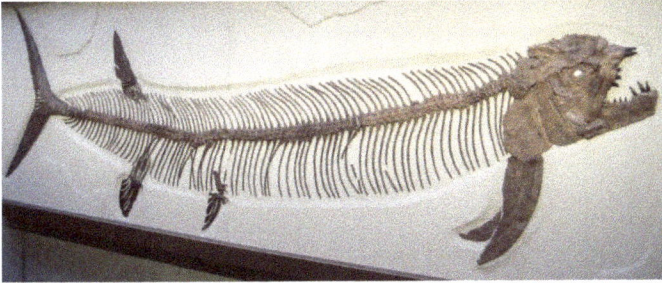

Xiphactinus *(zih-FACK-ti-nus)*, probably an extinct fish

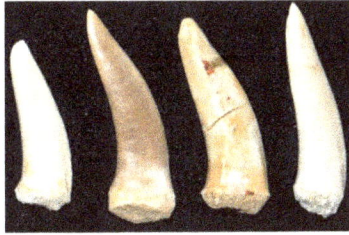

(Left) Enchodus *(EN-ko-dus)* – a nasty looking fish; (right) Encodus teeth

An exquisitely preserved fossil fish

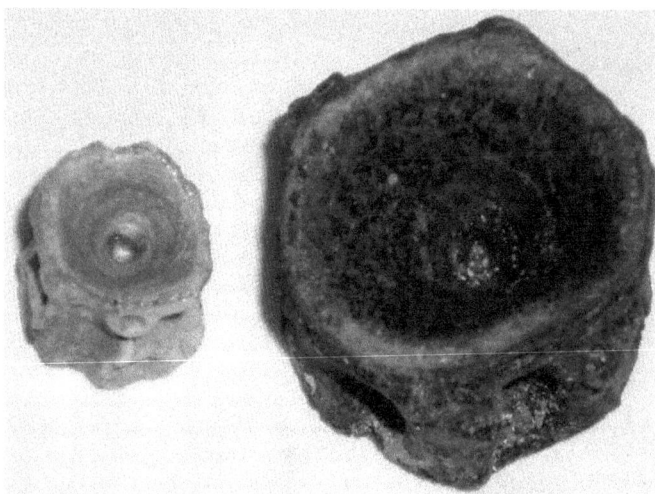

Fossil fish vertebrae, 1/2 inch and 1 ¼ inch across; generally they differ from shark vertebrae by being longer columns as opposed to flat disks (Florida)

Fossil shark vertebra, 2 ½ inches across; generally a flat disk (Florida)

Fossil Butte National Monument

Some of the world's best-preserved fossils are found in the flat-topped ridges of southwestern Wyoming. Fossilized fishes, insects, plants, reptiles, birds, and mammals are exceptional for their abundance, variety, and preservation. This is a real mix of Flood-based sediments all mixed together into a fine-grained limestone. They were probably deposited during the final days of the Flood as the mountains rose and the valleys sank down. Southwestern Wyoming, in particular, has a large deposit of sedimentary material rich in marine fossils.

Fossil Butte National Monument

Fossil Butte National Monument is located near the small Wyoming town of Kemmerer, Wyoming. While it is illegal to collect fossils inside the Monument, there are numerous private land owners who are more than eager to charge you a small fee for letting you look for fossils on their property. It is well worth the money, time and effort!

Fossil Butte National Monument is part of a larger formation called, The Green River Formation.

The most common fossil fish found at Fossil Butte National Monument

Fossil perch found at Fossil Butte National Monument

(Left) Fossil bird; (right) Fossil algae (Green River Formation, Wyoming)

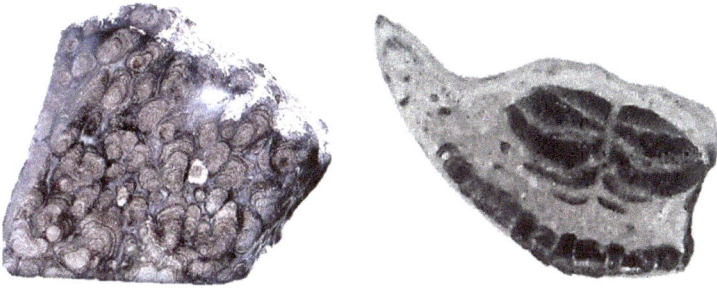

(Left) Fossil algae. two inches wide (northern Wyoming); (right) Fossil
Puffer Fish mouth part, 1 ½ inches across (Florida)

The extant dugong or manatee is also known as a sea cow. It is a sea
mammal.

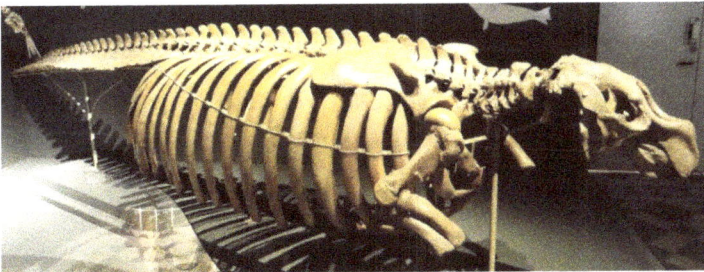

Dugong rib bones are prolific fossils and found in abundance in the rivers
of and off the coast of Florida. The rib bones look like thick, solid tubes
of curved rock. It is easy to overlook them.

Fossil Dugong rib bone (seven inches long) (Florida)

The whales – belong to the class of marine mammals called cetacean *(suh-TAY-she-un)*. In ancient Greek, the word *ketos* - Latinized as *cetus* - denotes a large fish, a whale, a shark, or a sea monster.

- Cetaceans have lungs, meaning they're air-breathers. The amount of time an individual can last without a breath varies from a few minutes to over two hours depending on the species.
- Cetaceans have especially powerful hearts. Likewise, the oxygen in the blood is distributed very effectively throughout the body.
- Cetaceans are warm-blooded animals. They hold a nearly constant, independent body temperature. Cold-blooded animals do not.
- Cetaceans give birth to fully developed calves and nurse them with high-fat milk from specific mammary glands. The embryonic development takes place in the body of the mother. During this time, the embryo is fed by a special nutritive tissue, the placenta.

The cetacea include whales, dolphins, and porpoises. Their fossils are actually quite common and not expensive to purchase. They were created on Day Five of Creation.

Then God said, "Let the waters teem with swarms of living creatures, and let birds fly above the earth in the open expanse of the heavens." God created the great sea monsters and every living creature that moves, with which the waters swarmed after their kind, and every winged bird after its kind; and God saw that it was good. God blessed them, saying, "Be fruitful and multiply, and fill the waters in the seas, and let birds multiply on the earth." There was evening and there was morning, a fifth day. Genesis 1:20-23

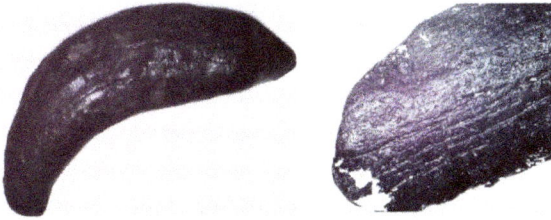

(Left) Fossil cetacean tooth, two inches long; (right) Fossil whale bone, about six inches long (Florida)

The rays – although many think it is a shark, it is generally grouped into its own category of cartilaginous fish called batoidea *(buh-TOY-dee-uh)*, made up of two words (*batis + oidea*), Latin and Greek, meaning flat fish. There are extinct varieties and extant varieties. And their fossils are quite common in certain locations. Florida is generally considered to be a great place to find their fossils, including the interior rivers.

(Left) Short-tail stingray ; (right) Eagle ray

89

Electric ray

Sawfish

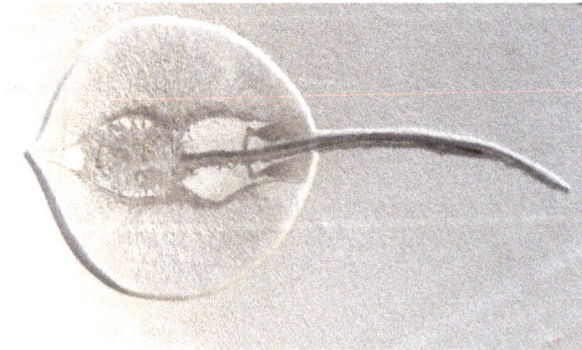

Ray fossil (Green River Formation, Wyoming)

Fossil ray tooth, about 1½ inches long (Florida)

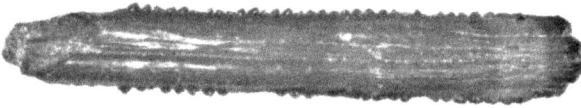

Ray fossil barbed tails. from two to four inches long (Florida)

IX: Identifying the
Vertebrate Land Fossils

 Fossils of land vertebrate animals are common, but their skeletons are not. There are plenty of fossil bones, but very few fossil skeletons. Concerning dinosaurs, one paleontologist estimated that there are about, "…311 skeletal fragments of dinosaurs worldwide for the last nine million years of the Cretaceous Period."[7] Of course this is the most significant period for the big dinosaurs. If his record of time was true, there certainly should be many more skeletons or partial skeletons available. Nine million years is a long, long time. The lack of fossilization cannot be the reason, as we find literally billions of fossil bones – everywhere! The picture is no better for the smaller dinosaurs and mammals. The Genesis Flood is the only reason why we see such a massive amount of disarticulated bones and very few skeletons.

Diplodocus *(dih-PLAH-doh-cus)* foot cast (Utah State Museum of Natural History)

---

[7] Dale Russell, *Hunting Dinosaurs* (Random House, New York, 1994), 258.

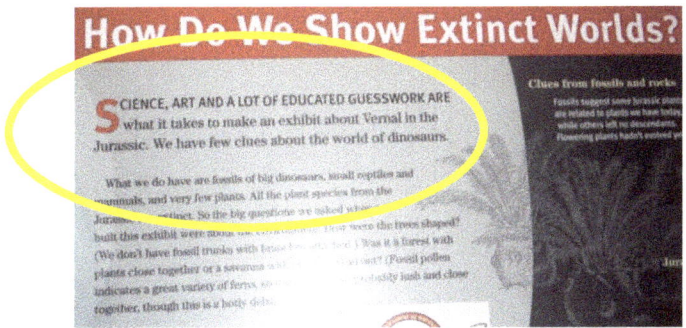

How Do We Show Extinct Worlds?

SCIENCE, ART AND A LOT OF EDUCATED GUESSWORK ARE what it takes to make an exhibit about Vernal in the Jurassic. We have few clues about the world of dinosaurs.

What we do have are fossils of big dinosaurs, small reptiles and mammals, and very few plants. All the plant species from the Jurassic are extinct. So the big questions are asked when built this exhibit were about the environment. How were the trees shaped? (We don't have fossil trunks with leaves [...]) Was it a forest with plants close together or a savanna [...] ? (Fossil pollen indicates a great variety of ferns, [...] [...] lush and close together, though this is a hotly deb[...]

Clues from fossils and rocks

Fossils suggest some Jurassic plants are related to plants we have today while others left no descendants. Flowering plants hadn't evolved ye[...]

As this sign from the Utah State Museum of Natural History so clearly reminds us, the three major components that go into reconstructing the past are science, art and a lot of educated guesswork. The sign goes on to state that we have very few actual skeletal fossils of dinosaurs, reptiles, mammals and plants. What is left out of these components is the Genesis Flood that would have the best explanation for why we have lots of fossil bones, but very few articulated skeletons.

Disarticulated dinosaur bones, two to four inches long, are the most abundant kind of dinosaur fossil (Hell Creek Formation in South Dakota).

A very nice and unusual dinosaur find, and characteristically disarticulated
(Utah)

Astounding Preservation

A Young Hadrosaur

These are the actual bones of a dinosaur excavated on the Monument in 2000 and 2001. Displayed here are most of the tail bones, the left pelvis, and portions of the left leg. 72 million years ago, the living animal was a 25-foot-long adolescent similar in shape to the *Parasaurolophus* in the courtyard, but without the tube-like crest.

Preservation of this animal is among the best known anywhere. The bones and delicate tendons are still in their connected positions. Amazingly, impressions of the skin were found still attached to the skeleton. Site evidence indicates that the complete animal was fossilized, but much of its body was lost through erosion and illegal collection. Specimens like these are extraordinary sources of information about these magnificent giants.

Paleontologists continue to make such discoveries in the Monument, assuring its place of prominence in our quest to understand the animals and events of the Late Cretaceous.

Even secular paleontologists acknowledge that even partial skeletal finds
such as these bones of a hadrosaur are amazingly preserved and very rare,
as seen in this museum display.

Miscellaneous dinosaur bones ranging from 1¾ inches to ½ inch long; fossil pieces like these take a keen eye. They are often over looked in the field (eastern Montana).

Dinosaur National Monument

Dinosaur National Monument is spread over 210,000 acres along the northwestern Colorado and northeastern Utah border. It is a bit out of the way, but if you want to see exposed dinosaur bones in their original matrix, this is the place to see them – well worth the drive! The following picture was taken at Dinosaur National Monument, Colorado . Finds like this are extremely rare and require huge amounts of money and man-hours to excavate. Many full and partial dinosaur skeletons have come from this wall of approximately 1,500 disarticulated dinosaur bones originally discovered in the 1800s. But those days seem to be over. It is just too expensive to extract the bones from these massive quarries. But these dinosaur remains preserved in this large rock wall are also a blessing for those of us who want to see the effects of the Genesis Flood.

Dinosaur National Monument

Many dinosaurs were huge! And it is these beasts that fired the imagination for movies like Jurassic Park. Pictured behind the T. rex replica is the Museum of the Rockies. One-time curator, Jack Horner, was the paleontological advisor for the Jurassic Park movies.

The Morrison Formation and the Hell Creek Formation

The Morrison Formation is centered in Wyoming and Colorado, with outcrops in Montana, North Dakota, South Dakota, Nebraska, Kansas, the panhandles of Oklahoma and Texas, New Mexico, Arizona, Utah, and Idaho. Equivalent rocks under different names are found in Canada.

The Morrison Formation was named after Morrison, Colorado, where Arthur Lakes discovered the first dinosaur fossils in the formation in 1877. That same year, it became the center of the Bone Wars, a fossil-collecting rivalry between early paleontologists Othniel Charles Marsh and Edward Drinker Cope.

The Morrison Formation is where most of the dinosaur fossils have been found in the U.S. And this makes sense, as it is essentially a washout of sediments that had covered the future Rocky Mountains. As the Rocky Mountains rose at the end of the Flood, miles of thick semi-wet sediments poured across the plains leaving a mess of disarticulated dinosaur bones, plants and many other types of fossils to the east and west of the Rockies. The formation covers an area of 600,000 square miles, although only a tiny fraction is exposed and accessible to geologists and paleontologists. Over 75% is still buried under the prairie! That in itself is big evidence of a watery/muddy catastrophe during the Genesis Flood.

Some of the most common fossils of the Morrison Formation are bits and pieces of fossil frill bone. Watch for the curvature of the bone as well as impressions, almost looking like canals, in the bone where blood vessels once existed.

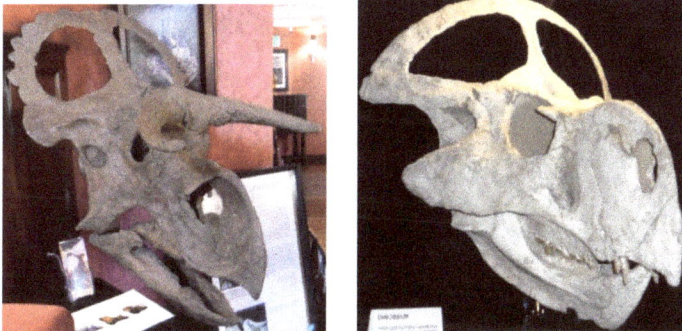

Here are two varieties of ceratopsian *(sair-uh-TOP-see-un)* dinosaurs, (left) one with horns (Utah) and (right) without horns (Mongolia); the shield above the horns is called the frill bone. This bone is often found in the Morrison Formation as disarticulated pieces.

One of the most iconic symbols of the dinosaurs is Triceratops *(try-SAIR-uh-tops)*. Once thought to be an individual species, it is now thought to be a juvenile version of Torosaurus *(TORE-uh-saur-us)*. Notice the shield (frill bone) behind the horns.

Lying on top of the Morrison Formation in eastern Montana is a formation known as the Hell Creek Formation. The formation is named after Hell Creek in northeast Montana. (Many geologists say, jokingly, that it got its name from the mass of disarticulated and highly eroded dinosaur bones.) If you want to find dinosaur bones in Montana, this is the place. You can expect to find bones from T. rex, triceratops, and hadrosaurs. There are several ranchers that will charge you a fee to dig on their property. Be sure to ask permission!

Fossil frill bone of Triceratops: each piece measures about four inches long. The channels you see in the bone are where the blood vessels were located within the bone (Hell Creek Formation).

(Left) Fossil frill bone of Triceratops, about three inches long; (right) Teeth of Triceratops; each tooth is 1 ½ inches long. (Hell Creek Formation).

Hadrosaur *(HAD-ruh-saur)*, or duck bill, dinosaur eggs: Most of the time you will have to purchase dinosaur eggs to have them in your collection. They are obtainable through auctions. But, buyer beware! There is a lot of fraud out there. (China)

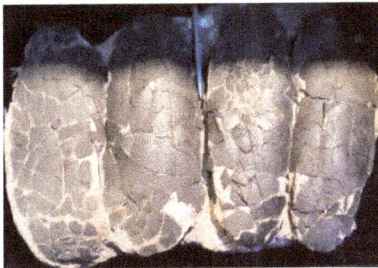

Intact dinosaur eggs – when searching known dinosaur fossil areas, keep your eyes open for shards of dinosaur egg. You can see by the picture that these eggs can be fragile and the chips of dinosaur eggshells, scattered about in the clay and sandstone can often be overlooked.

Dinosaur egg shell shard, about 1½ inches long; notice the perforated surface of the shell, a telltale sign that you have found part of an egg shell; these are easily missed in the field. (Argentina)

Disarticulated dinosaur bones, from two inches to five inches in length (Hell Creek Formation, eastern Montana)

Disarticulated dinosaur bones (Hell Creek Formation, South Dakota); the difference in color is due to the iron content of the sandstone in which they were found. Each piece measures about two inches in length.

A close-up view of the cross section of dinosaur bone showing its internal structure: specimen is about six inches across. (Utah)

An example of a dinosaur trace or ichno fossil is a gastrolith, meaning, stomach stone. These are often overlooked in the field. In areas known for their dinosaur bones, keep your eye open for polished stones that look different from other rocks surrounding them.

Gastroliths *(GAS-tro-liths)* or gizzard stones, about 1½ inches long; I found these in Montana near Ekalaka, where Triceratops fossils have been found.

Fossil small dinosaur rib bones about two inches long, and fossil small dinosaur digit bones about ¾ to one inch long – eastern Montana

Dinosaur vertebrae both are about three inches long (south central Montana)

Dinosaur coprolite *(KAH-pruh-lites)*, or dung or fossil poop, about four inches long (Utah)

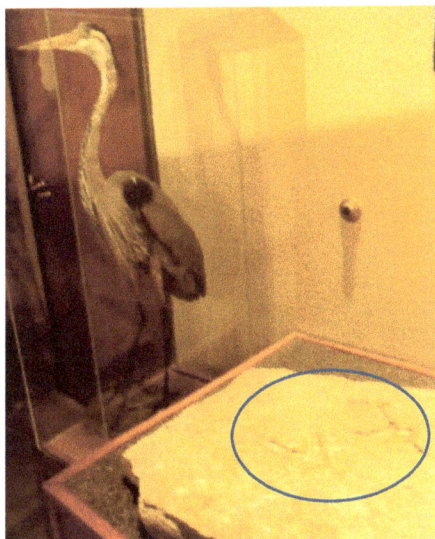

Many times, fossils of modern animals and plants are found in dinosaur rock layers. This is what we would expect, as creationists. It is interesting that some secular museums show modern animals and plants in their dinosaur displays, contradictory to their own beliefs! Notice the stork tracks in this display that are found in dinosaur rock layers! (Fruita, CO)

(Left) Fossil turtle shell about 10 inches long – turtles are a geological anomaly! Paleontologists tell us that the dinosaurs died out 65 million years ago, yet turtles are found in dinosaur rock layers looking no different than their modern counterparts.
(Right) Disarticulated turtle parts about one inch in length (Montana)

Fossil turtle egg about 1 ½ inch long (South Dakota)

Three other animals that survived the so-called dinosaur extinction of 65 million years ago were the crocodile, alligator, and snake, looking no different than their modern counterparts.

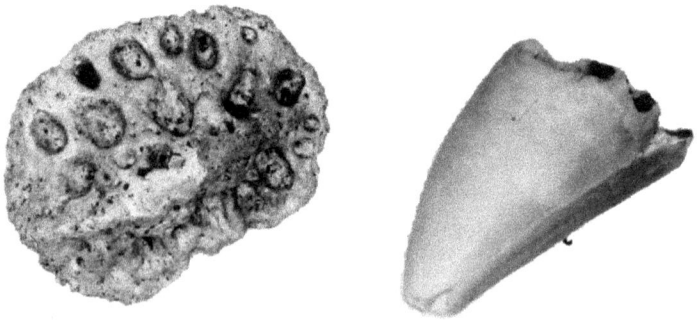

(Left) Fossil crocodile scute (armor plate) about 1¾ inches long; (right) Fossil alligator tooth about one inch long

(Top) Fossil alligator vertebra; (bottom) Fossil snake vertebrae

Using our Basic Biblical Framework introduced in section V, we now come to The Post-Flood period. Please note: we do not know for sure whether some of the mammal fossils we find were actually a part of the Genesis Flood or of The Post Flood Period that brought on the great ice event. But there would have been animals that had survived the great Genesis Flood by being saved by the ark of Noah. These animals, mammals, and dinosaurs alike, would have come off of the ark and then migrated and reproduced according their kinds. In fact these animals had at least a hundred year advance on mankind, as man chose to stay in the Mesopotamian region in direct violation of God's command to spread out and multiply and fill the earth.

In any event the period that initially followed the Genesis Flood would have been another type of global catastrophic event. Major volcanic eruptions and tectonic uplift would have lasted for some time as the earth began to recover from that awful watery catastrophe. One of these resultant catastrophes was most likely an ice event, called by modern geologists, The

Ice Age. This significant period probably lasting upwards of 500-700 years would have been responsible for eventually cutting off many of our modern continents and causing massive extinction among many mammals including the mammoth, the ancient bison, saber-toothed cat, wooly rhinoceros, and many others. It could also account for why the dinosaurs may never have really thrived after the Flood.

Some of the animals typically identified with The Ice Age are pictured below. Remember, though, some of the fossils labeled ice age could be remnants of the Genesis Flood. We cannot know for sure in some cases. Geologists teach that the big mammals did not evolve until after the dinosaurs disappeared. But of course we know from Genesis 1 that both mammals and dinosaurs were created on Day Six of Creation week. So, many of the fossils we find could be a real mix of both mammals and dinosaurs.

But we also remember that there was significant global change going on after the Flood that could have provided ideal conditions in some areas for significant fossilization.

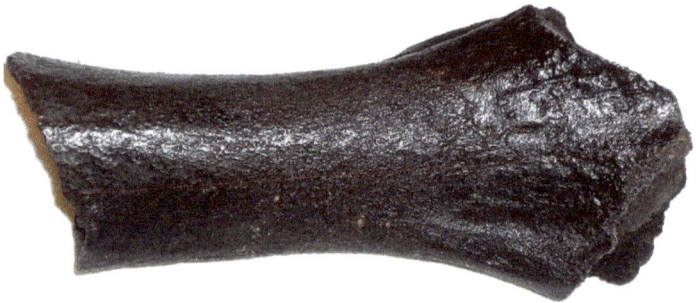

Fossil mammal bone – most likely that of a bison, about eight inches long

The mammal bones are easily recognizable in that they are generally hollow, characteristic of mammal bones. The dark color is due to the presence of phosphorus. The phosphorus beds of Florida are filled with many such dark colored bones of all sorts of vertebrate fossils.

Other mammal fossils found together at the La Brea tar pits in California

(Left) Fossil beaver tooth (1¾ inches long)
(Right) Teeth of a small camel, 1¾ inches long

(Left) Complete fossil mammoth tusks, about five feet long, are extremely rare, but are readily available in recognizable disarticulated pieces. (Right) Mammoth tooth, about one foot long

The mammoth skeleton – from a composite of fossil bones found at La Brea Tar Pits, California

Fossil mammoth bone, about four inches long;

Fossil tusk – disarticulated pieces are common, about four inches long

Disarticulated mammoth tusk, 13 inches long (Holland)

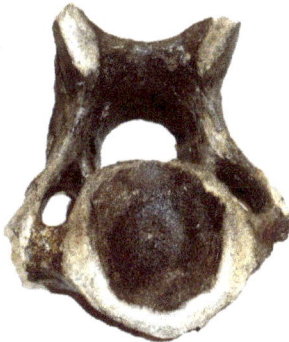

Mammoth vertebra, about seven inches long (Holland)

The Mastodon *(MAS-tuh-don)* – composite of fossils (La Brea Tar Pits, California)

(Left) Fossil mastodon partial tooth, about 2 ½ inches long (Florida); (right) Fossil mastodon tooth, about five inches long (Holland)

The saber-toothed cat – composite of fossils (La Brea Tar Pits, California)

Fossil horse skeleton – composite of fossils (La Brea Tar Pits, California)

Fossil horse tooth, about two inches long (Florida)

(Left) Fossil deer antler, about three inches long; (right) Fossil mammal
bone about three inches long (Florida);
notice the hollow nature of the bone.

(Left) Fossil small mammal vertebra, about one inch long (Nebraska)
(Right) Fossil mammal bone, about four inches long (Florida)

Fossil mammal coprolite (poop!), about 1 ½ inches long (Nebraska);
fossil poops are not uncommon!

Large mammal vertebra, about four inches long (Florida)

X. Other Fossils

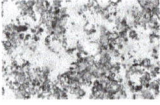

Let's take a look at the rarer and often more expensive fossils to obtain. These fossils would include insects, plants, and microfossils.

Insects

(Left) Fossil cricket in limestone (Wyoming) ; (right) Ant in amber (Baltic Sea)

(Left) Ant in limestone (Wyoming); (right) Wasp in limestone (Wyoming)

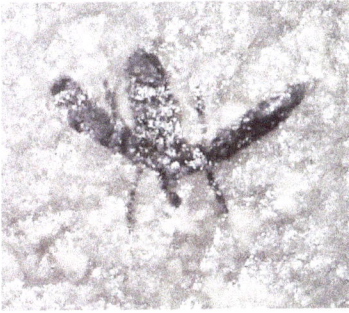

(Left) Bee in limestone (Wyoming); (right) Fly in amber (Baltic Sea)

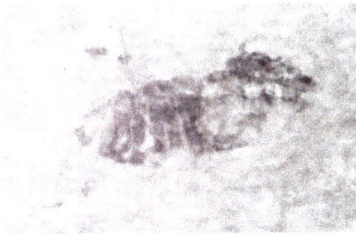

(Left) Cricket in amber (Baltic Sea) ; (right) Fossil larva in limestone (Wyoming)

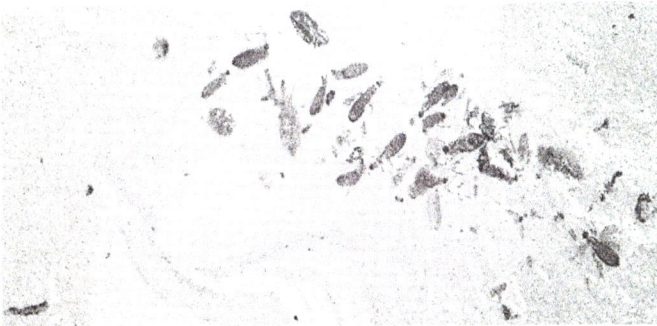

Insect swarm in limestone (Wyoming)

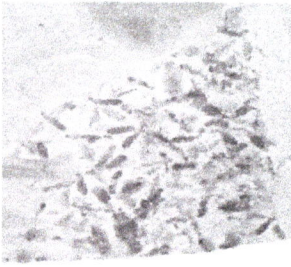

(Left) Insect swarm in limestone (Wyoming), (right) Mosquito in amber (Baltic Sea)

Fossil wasp in limestone (Wyoming)

Plants

Ferns are readily available as fossils, as is petrified wood. I used to walk up and down the railroad tracks laid on shale in an area of Montana where I grew up. I would search for these fossil ferns imprinted on the shale pieces. In addition to fossil ferns, you can also look for petrified wood, fossil pine cones, and other assorted plants.

(Left) Fossil ferns in mudstone (Illinois); (right) Fossil reed (Illinois)

Fossil pine cones each, about one inch long

Incredible as it may seem, these are real fossils of palm branches in limestone from the Green River Formation in Wyoming. The fossils in the picture on my right side are fish along with water plants, also in limestone, from Wyoming.

In this close-up picture of the palm branch, you can see some fossil fish in the same limestone indicating a watery burial for these palm branches!

In this picture, you can see some varieties of petrified wood. In the upper left is petrified wood with insect borings. The greenish piece in the middle is opalized wood. Opal is a form of silica that has been altered by hot water. It is quite common in Washington and parts of Oregon. The various pieces of polished wood are from Arizona where iron rich sediments have contributed to the wood's rich red color.

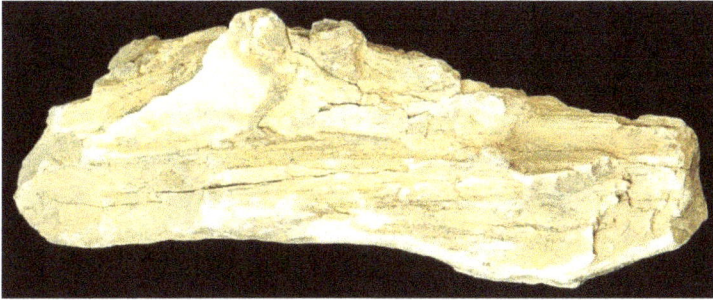

Petrified wood with the help of volcanic ash from Yellowstone country on the east side of Yellowstone National Park

Author standing next to a huge petrified tree; Notice two things: there is no root ball at the end of the tree indicating the tree is not in growth location; and notice the planation surfaces in the background. These geologic landforms indicate a great body of water that once planed this area, leaving behind trees that were either petrified in the process, or had been previously petrified in another location, and washed in.

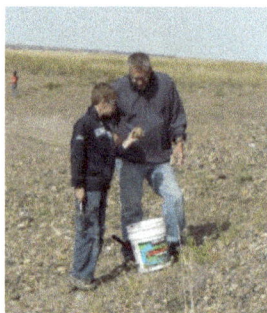

This gravel pit near Glendive, Montana was deposited during the great ice event dam breach of Glacial Lake Great Falls. These kinds of gravel pits are treasure troves for finding petrified wood and other fossils.

Fossil leaves from Utah at the Utah State Museum – the sign reads Familiar Leaves from the Past, indicating no evolutionary change over millions of years!

Microfossils

Microfossils are so small that you need a magnifying glass to observe them. It is a specialty within paleontology and takes a great deal of patience to study them. They are not initially very exciting. In case you are wondering though, the following is what a microfossil looks like.

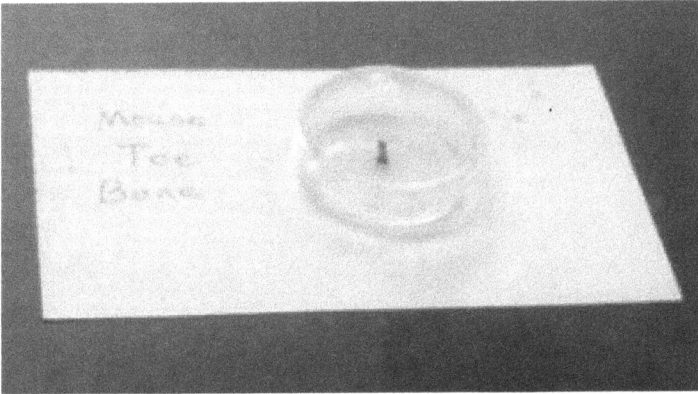

Pictured here is a very tiny fossil of a mouse toe bone, approximately 1/8" long. It takes a very keen interest and eye to study these kinds of fossils. (La Brea Tar Pits, California)

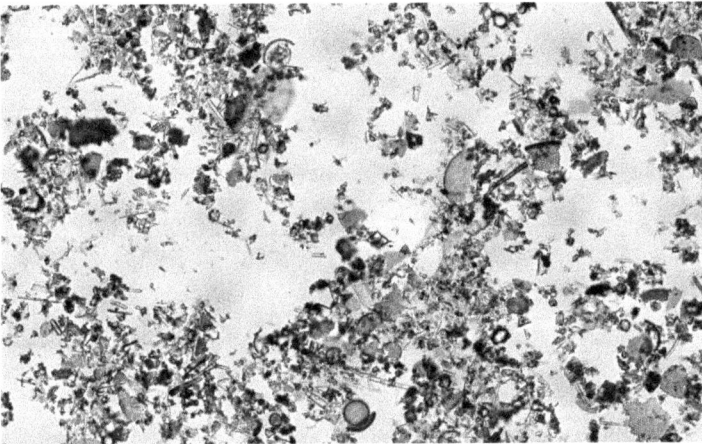

Diatomaceous *(DIE-uh-toe-MAY-she-us)* earth – the matrix within which you find diatoms (tiny animals with skeletons of calcite)

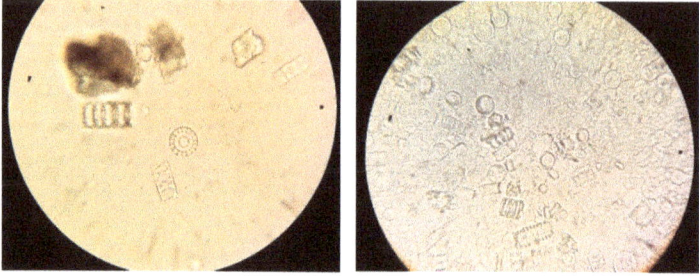

Fossil diatoms in diatomaceous earth under a microscope

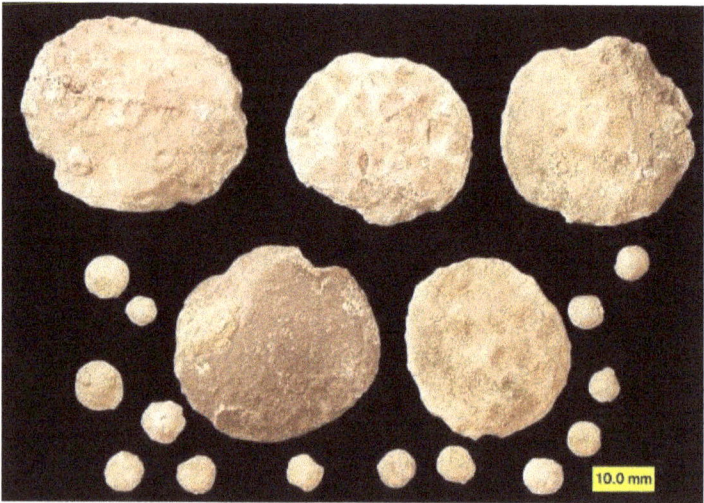

Nummulitids *(noo-MOO-lih-tids)*, or fossil protozoa

(Left) Chert, a siliceous sedimentary rock, contains bacteria and cyanobacteria *(si-AN-o-back-TIH-ree-uh)* fossils. It can be either light or dark. When it is dark, it is called flint. (Right) This chalk contains the calciferous skeletons of diatoms.

Conodonts

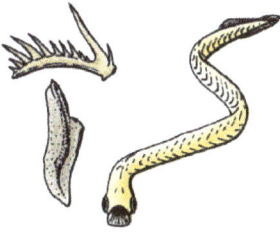

Conodonts *(CONE-uh-dahnt)* – from the Greek kōnos (cone), and odont (tooth); they are extinct chordates resembling eels, classified in the class Conodonta. For many years, they were known only from tooth-like microfossils found in isolation like the ones above. They are now called conodont elements and are recognized as part of the creature we know as a conodont. Knowledge about their soft tissues remains limited. Chordates are animals possessing a notochord, a hollow dorsal nerve cord, pharyngeal slits, an endostyle, and a post-anal tail for at least some period of their life cycles.

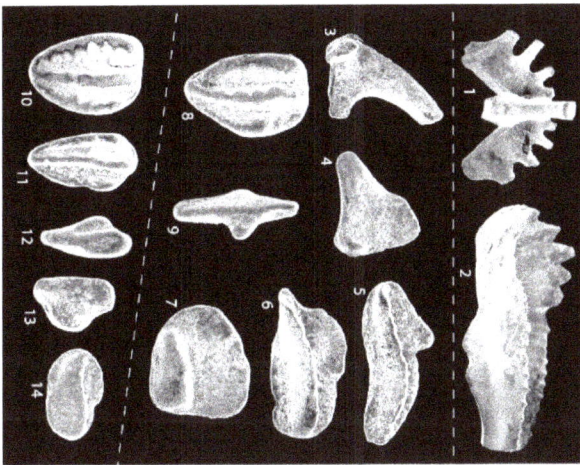

Conodonts

Notes

Photo Credits

I.
Fossil ammonite: From author's collection, photo by Patrick Nurre, 4. Field pictures: by Vicki Nurre, 6-7. Guide books: photo by Vicki Nurre, 7.

II.
Cricket in amber: From author's collection, photo by Patrick Nurre, 9.

III.
Fossil mastodon partial tooth: From author's collection, photos by Patrick Nurre, 15.
Cauditeryx: "Caudipteryx zoui - Untere Kreide - Liaoning-China" by Ra'ike (see also: de:Benutzer:Ra'ike) - Own work. Licensed under CC BY-SA 3.0 via Commons - https://commons.wikimedia.org/wiki/File:Caudipteryx_zoui_-_Untere_Kreide_-_Liaoning-China.jpg#/media/File:Caudipteryx_zoui_-_Untere_Kreide_-_Liaoning-China.jpg, 16. Fossil Coelacanth: "Undina penicillata" by User:Haplochromis - Self-photographed. Licensed under CC BY-SA 3.0 via Commons - https://commons.wikimedia.org/wiki/File:Undina_penicillata.jpg#/media/File:Undina_penicillata.jpg, 19. Reconstruction of living coelacanth: "Latimeria chalumnae replica" by © Citron /. Licensed under CC BY-SA 3.0 via Commons - https://commons.wikimedia.org/wiki/File:Latimeria_chalumnae_replica.jpg#/media/File:Latimeria_chalumnae_replica.jpg, 19. Baculite pieces: "SouthDakotaBaculites" by DanielCD - Own work. Licensed under CC BY-SA 3.0 via Commons - https://commons.wikimedia.org/wiki/File:SouthDakotaBaculites.jpg#/media/File:SouthDakotaBaculites.jpg, 20. Scaphopod: "Antalis vulgaris" by © Hans Hillewaert. Licensed under CC BY-SA 4.0 via Commons - https://commons.wikimedia.org/wiki/File:Antalis_vulgaris.jpg#/media/File:Antalis_vulgaris.jpg. 20. Trilobite: "BLW Trilobite (Paradoxides sp.)" by Mike Peel. Licensed under CC BY-SA 2.0 uk via Commons https://commons.wikimedia.org/wiki/File:BLW_Trilobite_(Paradoxides_sp.).jpg#/media/File:BLW_Trilobite_(Paradoxides_sp.).jpg, 21.

IV.
Xiphactinus: Xiphactinus audax fossil" by Eduard Solà Vázquez - Own work. Licensed under CC BY-SA 3.0 via Commons - https://commons.wikimedia.org/wiki/File:Xiphactinus_audax_fossil.jpg#/media/File:Xiphactinus_audax_fossil.jpg, 22. Fossil in amber: "Leptofoenus pittfieldae (male) rotated" by Leptofoenus_pittfieldae_(male).JPG: Michael S. Engelderivative work: Kevmin (talk) - The first fossil leptofoenine wasp (Hymenoptera, Pteromalidae): A new species of Leptofoenus in Miocene amber from the Dominican Republic doi:10.3897/zookeys.13.159, via Leptofoenus_pittfieldae_(male).JPG. Licensed under CC BY 3.0 via Commons - https://commons.wikimedia.org/wiki/File:Leptofoenus_pittfieldae_(male)_rotated.JPG#/media/File:Leptofoenus_pittfieldae_(male)_rotated.JPG, 23. Fossil Fish: "Priscacara liops Green River Formation" by Didier Descouens - Own work. Licensed under CC BY-SA 4.0 via Commons - https://commons.wikimedia.org/wiki/File:Priscacara_liops_Green_River_Formation.jpg#/media/File:Priscacara_liops_Green_River_Formation.jpg, 23.Petrified Wood: Photo by Vicki Nurre, 23. Fossil preserved in Shale: "Opabinia smithsonian" by Original uploader was Jstuby at en.wikipedia - Transferred from en.wikipedia; transferred to Commons by User:FunkMonk using CommonsHelper.. Licensed under Public Domain via Commons - https://commons.wikimedia.org/wiki/File:Opabinia_smithsonian.JPG#/media/File:Opabinia_smithsonian.JPG, 24. Dinosaurs bones in sandstone: "Dinosaur National Monument-inside the Dinosaur Quarry building" by Transferred from en.wikipedia to Commons.; originally from http://www.cr.nps.gov/museum/treasures/html/Q/h020.html. Licensed under Public Domain via Commons - https://commons.wikimedia.org/wiki/File:Dinosaur_National_Monument-inside_the_Dinosaur_Quarry_building.jpeg#/media/File:Dinosaur_National_Monument-inside_the_Dinosaur_Quarry_building.jpeg, 24.

V.
Fossil algae: "Stromatolite Fossil from Wyoming" by James St. John (jsj1771) http://www.flickr.com/people/jsjgeology/ -

126

http://www.flickr.com/photos/jsjgeology/8362709905/sizes/o/in/set-72157632477572336/. Licensed under CC BY 2.0 via Commons - https://commons.wikimedia.org/wiki/File:Stromatolite_Fossil_from_Wyoming.jpg#/media/File:Stromatolite_Fossil_from_Wyoming.jpg, 25. Fossil clam: From author's collection, photo by Patrick Nurre, 26. Fossil skull: photo by Patrick Nurre, 32. Fossil turtle shell: From author's collection, photo by Patrick Nurre, 26. Fossil dinosaur bones: From author's collection, photo by Patrick Nurre, 32. Fossil insect: From author's collection, photo by Patrick Nurre, 26. Agatized wood: Photo by Patrick Nurre, 26. Fossil gastropods: From author's collection, photo by Patrick Nurre, 27. Fossil ammonite: From author's collection, photo by Patrick Nurre, 27. Fossil ammonites: From author's collection, photo by Patrick Nurre, 27. Starfish cast: From author's collection, photo by Patrick Nurre, 28. Dinosaur track cast: Photo by Patrick Nurre, 28. Fern cast: From author's collection, photo by Patrick Nurre, 28. Trilobite mold and cast: Photo by Oscar Sanchez, Bolivian fossils, used by permission, 29. Dinosaur track ichno fossil: Photo by Patrick Nurre, 29. Fossil worm tubes: From author's collection, photo by Patrick Nurre, 30. Fossils shrimp burrow: From author's collection, photo by Patrick Nurre, 30. Worm burrows ichno fossils: From author's collection, photo by Patrick Nurre, 30. Trilobite tracks: From author's collection, photo by Patrick Nurre, 30. Fossil worms in limestone: Photo by Heidi Noggle, used by permission, 31. Insect in amber: "Amber2" by Anders L. Damgaard - www.amber-inclusions.dk - Baltic-amber-beetle. Original uploader was AmericanXplorer13 at en.wikipedia - This file was derived from: Baltic amber inclusions - Ant (Hymenoptera, Formicidae)8.JPGTransferred from en.wikipedia; transferred to Commons by User:Common Good using CommonsHelper.(Original text : Work of Baltic-amber-beetle). Licensed under CC BY-SA 3.0 via Commons - https://commons.wikimedia.org/wiki/File:Amber2.jpg#/media/File:Amber2.jpg, 31. Bituminour coal: From author's personal collection, photo by Heidi Nobble, used by permission, 32. Sydney mines: "Sydney Mines Point Aconi Seam 038" by Rygel, M.C. - Own work. Licensed under CC BY-SA 3.0 via Commons - https://commons.wikimedia.org/wiki/File:Sydney_Mines_Point_Aconi_Seam_038.JPG#/media/File:Sydney_Mines_Point_Aconi_Seam_038.JPG, 32. Botryoidal mass of pyrolusite: "Pyrolusite botryoidal" by Aram Dulyan (User:Aramgutang) - Own work. Licensed under Public Domain via Commons - https://commons.wikimedia.org/wiki/File:Pyrolusite_botryoidal.jpg#/media/File:Pyrolusite_botryoidal.jpg, 33. Pyrolusite: From author's collection, photo by Heidi Noggle, used by permission, 33. Pyrolusite up close: Photo by Patrick Nurre, 33. Large concretion: Photo by Vicki Nurre, 34. Concretions: "The shapes of concretions 1" by Brocken Inaglory - Own work. Licensed under GFDL via Commons - https://commons.wikimedia.org/wiki/File:The_shapes_of_concretions_1.JPG#/media/File:The_shapes_of_concretions_1.JPG, 34. Moqui marbles: "MoquiMarble1" by Paul Heinrich - Own work. Licensed under CC BY 3.0 via Commons - https://commons.wikimedia.org/wiki/File:MoquiMarble1.jpg#/media/File:MoquiMarble1.jpg, 34. Sandstone concretions: Photo courtesy of Tim and Candey (earths.ancient.gifts), 34. Fossil crab: From author's collection, photo by Patrick Nurre, 35. Scaphopod: From author's collection, photo by Patrick Nurre, 35. Mud cracks: From author's collection, photo by Patrick Nurre, 36.Fossil ripple marks: Photo by Patrick Nurre, 37.

VI.
Agatized wood: Photo by Patrick Nurre. 38.Petrified Forest: Photo by Vicki Nurre, 38. Sandstone, Petrified Forest: Photo by Patrick Nurre,39. Conglomerate sediments: Photo by Patrick Nurre, 39. Petrified logs: Photo by Patrick Nurre, 40. Rounded stones: Photo by Patrick Nurre, 40. Petrified log and rounded stones: Photo by Patrick Nurre, 41. Petrified log: Photo by Patrick Nurre, 41. Close-up of petrified log: Photo by Patrick Nurre, 41. Diagrams: Patrick Nurre, 43.

VII.
Crinoid: "Fossile-seelilie" by Berengi - Transferred from de.wikipedia to Commons.. Licensed under CC BY-SA 3.0 via Commons - https://commons.wikimedia.org/wiki/File:Fossile-seelilie.jpg#/media/File:Fossile-seelilie.jpg, 47.
Four photos of limestone with and without fossils: From author's collection, photo by Patrick Nurre, 48. Coquina: From author's collection, photo by Patrick Nurre, 49.
Sample shales: From author's collection, photo by Patrick Nurre, 49. Shale formation: Photo by Patrick Nurre, 49. Fossil fern: From author's collection, photo by Patrick Nurre, 50. Sandstone formation: Photo by Patrick Nurre, 50. Ripples in sandstone: From author's personal collection, photo by Patrick Nurre, 50. Tracks in sandstone: From author's personal collection, photo by Patrick Nurre, 51. Kainus invius: "Kainops invius lateral and ventral" by Moussa Direct Ltd. - Moussa Direct Ltd. image archive. Licensed under CC BY-SA 3.0 via Commons - https://commons.wikimedia.org/wiki/File:Kainops_invius_lateral_and_ventral.JPG#/media/File:Kainops_invius_lateral_and_ventral.JPG, 52. Representative trilobites: From author's collection, photo by

Patrick Nurre, 52-53. Anatomy of live coral polyp: "Coral polyp" by NOAA - NOAA website. Licensed under Public Domain via Commons -
https://commons.wikimedia.org/wiki/File:Coral_polyp.jpg#/media/File:Coral_polyp.jpg, 54.
Representative fossil corals: From author's collection, photos by Patrick Nurre, 54-55. Crinoid pieces and fossils: From author's collection, photos by Patrick Nurre, 56. Crinoid: "Fossile-seelilie" by Berengi - Transferred from de.wikipedia to Commons.. Licensed under CC BY-SA 3.0 via Commons -
https://commons.wikimedia.org/wiki/File:Fossile-seelilie.jpg#/media/File:Fossile-seelilie.jpg, 56.
Actinocrinus: Actinocrinus indiana 330m" by Sanjay Acharya - Own work. Licensed under CC BY-SA 3.0 via Commons -
https://commons.wikimedia.org/wiki/File:Actinocrinus_indiana_330m.jpg#/media/File:Actinocrinus_indiana_330m.jpg, 56. Crinoids: "Crinoids iowa 330m" by Sanjay Acharya - Own work. Licensed under CC BY-SA 3.0 via Commons -
https://commons.wikimedia.org/wiki/File:Crinoids_iowa_330m.jpg#/media/File:Crinoids_iowa_330m.jpg, 57. Crinoid hold-fasts: "OrdCrinoidHoldfasts" by Wilson44691 - Own work. Licensed under Public Domain via Commons -
https://commons.wikimedia.org/wiki/File:OrdCrinoidHoldfasts.jpg#/media/File:OrdCrinoidHoldfasts.jpg, 57. Crinoid stem pieces: "JurassicCrinoidsIsrael" by Wilson44691 at English Wikipedia - Photograph taken by Mark A. Wilson (Department of Geology, The College of Wooster). [1]. Licensed under Public Domain via Commons -
https://commons.wikimedia.org/wiki/File:JurassicCrinoidsIsrael.JPG#/media/File:JurassicCrinoidsIsrael.JPG, 57. Living crinoid: "Crinoid on the reef of Batu Moncho Island" by Alexander Vasenin - Own work. Licensed under CC BY-SA 3.0 via Commons -
https://commons.wikimedia.org/wiki/File:Crinoid_on_the_reef_of_Batu_Moncho_Island.JPG#/media/File:Crinoid_on_the_reef_of_Batu_Moncho_Island.JPG, 57.
Living sea urchin diagram: "Urchin9b" by Alex Ries - Personal email message from artist, as derived from http://abiogenisis.deviantart.com/art/Sea-Urchin-Anatomy-271355683. Licensed under CC BY-SA 4.0 via Commons - https://commons.wikimedia.org/wiki/File:Urchin9b.jpg#/media/File:Urchin9b.jpg, 58.
Fossils sea urchin, spines and sand dollars: From author's personal collection, photo by Patrick Nurre, 58. Fossil Bryozoan: From author's personal collection, photo by Patrick Nurre, 59. Mortality plate: From author's personal collection, photo by Patrick Nurre, 59.

VIII.
Baculites: From author's collection, photos by Patrick Nurre, 60. Lingula anatina: "LingulaanatinaAA" by Wilson44691 - Own work. Licensed under Public Domain via Commons -
https://commons.wikimedia.org/wiki/File:LingulaanatinaAA.JPG#/media/File:LingulaanatinaAA.JPG, 60. Mortality plate: "Cincinnetina meeki (Miller, 1875) slab 2" by Wilson44691 - Own work. Licensed under CC0 via Commons -
https://commons.wikimedia.org/wiki/File:Cincinnetina_meeki_(Miller,_1875)_slab_2.jpg#/media/File:Cincinnetina_meeki_(Miller,_1875)_slab_2.jpg, 66==, 61.
Oil lamp: "Oil Lamp Christian Symbol" by No machine-readable author provided. Rama assumed (based on copyright claims). - No machine-readable source provided. Own work assumed (based on copyright claims).. Licensed under CC BY-SA 2.0 fr via Commons -
https://commons.wikimedia.org/wiki/File:Oil_Lamp_Christian_Symbol.jpg#/media/File:Oil_Lamp_Christian_Symbol.jpg, 61. Fossil brachiopods: From author's collection, photo by Patrick Nurre, 62. Giant clam: "Tridacna gigas.001 - Aquarium Finisterrae" by Drow_male - Own work. Licensed under GFDL via Commons - https://commons.wikimedia.org/wiki/File:Tridacna_gigas.001_-_Aquarium_Finisterrae.JPG#/media/File:Tridacna_gigas.001_-_Aquarium_Finisterrae.JPG, 63. Razor clams: "Ensis phaxoides-V" by Jan Johan ter Poorten; modified by Tom Meijer -
http://www.spirula.nl/images/nl_soorten/marien/Ensis_phaxoides.jpg. Licensed under CC BY-SA 3.0 via Commons - https://commons.wikimedia.org/wiki/File:Ensis_phaxoides-V.jpg#/media/File:Ensis_phaxoides-V.jpg, 64. Valve dorsal view: "Valve-DorsalView collored" by Valve-DorsalView.png: Muriel Gottropderivative work: Sundance Raphael (talk) - Valve-DorsalView.png. Licensed under CC BY-SA 1.0 via Commons - https://commons.wikimedia.org/wiki/File:Valve-DorsalView_collored.svg#/media/File:Valve-DorsalView_collored.svg, 64.
Fossil oyster: "GryphaeaCretaceousTexas" by Wilson44691 - Own work. Licensed under Public Domain via Commons -
https://commons.wikimedia.org/wiki/File:GryphaeaCretaceousTexas.jpg#/media/File:GryphaeaCretaceousTexas.jpg, 64. Fossil oyster shell: "Exogyra costata Prairie Bluff Fm Maastrichtian" by Wilson44691 - Own work. Licensed under CC0 via Commons -
https://commons.wikimedia.org/wiki/File:Exogyra_costata_Prairie_Bluff_Fm_Maastrichtian.JPG#/media/File:Exogyra_costata_Prairie_Bluff_Fm_Maastrichtian.JPG, 64.
Fossil oyster shell: From author's collection, photo by Patrick Nurre, 65. Fossil mussel: From author's collection, photo by Patrick Nurre, 65. Mortality plate: From author's collection, photo by Patrick Nurre, 75. Living gastropod diagram: "Snail diagram-en edit1" by Original by Al2, English captions and other

edits by Jeff Dahl - Own work. Licensed under GFDL via Commons -
https://commons.wikimedia.org/wiki/File:Snail_diagram-en_edit1.svg#/media/File:Snail_diagram-
en_edit1.svg, 66.
Fossil gastropod: "Turritellatricarinata" by Wilson44691 - Own work. Licensed under Public Domain via
Commons -
https://commons.wikimedia.org/wiki/File:Turritellatricarinata.jpg#/media/File:Turritellatricarinata.jpg,
66. Fossil gastropods: From author's collection, photos by Patrick Nurre, 66. Flat snail: From author's
collection, photo by Patrick Nurre, 66.
Fossil snails: From author's collection, photo by Patrick Nurre, 67. Fossil gastropod: From author's
collection, photos by Patrick Nurre, 67. Various mollusk fossils: From author's collection, photo by
Patrick Nurre, 68. Living nautilus: "Nautilus Palau" by Manuae - Own work. Licensed under CC BY-SA
3.0 via Commons -
https://commons.wikimedia.org/wiki/File:Nautilus_Palau.JPG#/media/File:Nautilus_Palau.JPG, 69.
Octopus: "Octopus vulgaris2" by Beckmannjan at the German language Wikipedia. Licensed under CC
BY-SA 3.0 via Commons -
https://commons.wikimedia.org/wiki/File:Octopus_vulgaris2.jpg#/media/File:Octopus_vulgaris2.jpg,
69. Squid: "Loligo vulgaris" by © Hans Hillewaert. Licensed under CC BY-SA 4.0 via Commons -
https://commons.wikimedia.org/wiki/File:Loligo_vulgaris.jpg#/media/File:Loligo_vulgaris.jpg, 69.
Common living squid: Photo © Hans Hillewaert, licensed under CC BY-SA 4.0, found at
https://en.wikipedia.org/wiki/Common_cuttlefish#/media/File:Sepia_officinalis_(aquarium).jpg, 69.
Fossil ammonites: From author's collection, photos by Patrick Nurre, 70. Baculites: From author's
collection, photos by Patrick Nurre, 71. Fossil cephalopods: From author's collection, photos by Patrick
Nurre, 71. Fossil graveyard: From author's collection, photo by Patrick Nurre, 72.
Crustacean structure: "Krillanatomykils" by Uwe Kils - english wikipedia, original upload 15 June 2005 by
en:User:Kils. Licensed under CC BY-SA 3.0 via Commons -
https://commons.wikimedia.org/wiki/File:Krillanatomykils.jpg#/media/File:Krillanatomykils.jpg, 73.
Fossil shrimp: From author's collection, photo by Patrick Nurre, 73.
Fossil lobster: "Eryma mandelslohi (Krebs) - Oberer Brauner Jura - Bissingen unter Teck" by Ra'ike (see
also: de:Benutzer:Ra'ike) - Own work. Licensed under CC BY-SA 3.0 via Commons -
https://commons.wikimedia.org/wiki/File:Eryma_mandelslohi_(Krebs)_-_Oberer_Brauner_Jura_-
_Bissingen_unter_Teck.jpg#/media/File:Eryma_mandelslohi_(Krebs)_-_Oberer_Brauner_Jura_-
_Bissingen_unter_Teck.jpg, 74. Fossil crab: Photo by Rene Sylvestersen, Licensed under CC BY-SA 3.0
via Commons, 74. Fossil barnacle: "Chesaconcavus top view" by Wilson44691 - Own work. Licensed
under CC0 via Commons -
https://commons.wikimedia.org/wiki/File:Chesaconcavus_top_view.jpg#/media/File:Chesaconcavus_t
op_view.jpg, 74. Living protozoa: "Mikrofoto.de-Blepharisma japonicum 15" by Frank Fox -
http://www.mikro-foto.de. Licensed under CC BY-SA 3.0 de via Commons -
https://commons.wikimedia.org/wiki/File:Mikrofoto.de-
Blepharisma_japonicum_15.jpg#/media/File:Mikrofoto.de-Blepharisma_japonicum_15.jpg, 75. Fossil
protozoa: From author's collection, photo by Patrick Nurre, 86. Fossil graveyard: From author's
collection, photo by Patrick Nurre, 75.

IX.
Xiphactinus: Xiphactinus audax fossil" by Eduard Solà Vázquez - Own work. Licensed under CC BY-SA
3.0 via Commons -
https://commons.wikimedia.org/wiki/File:Xiphactinus_audax_fossil.jpg#/media/File:Xiphactinus_auda
x_fossil.jpg, 76. Vertebral column: "718 Vertebra" by OpenStax College - Anatomy & Physiology,
Connexions Web site. http://cnx.org/content/col11496/1.6/, Jun 19, 2013. Licensed under CC BY 3.0
via Commons -
https://commons.wikimedia.org/wiki/File:718_Vertebra.jpg#/media/File:718_Vertebra.jpg, 76. Shark
teeth: "CretaceousSharkTeeth061812" by Wilson44691 - Own work. Licensed under CC BY-SA 3.0 via
Commons -
https://commons.wikimedia.org/wiki/File:CretaceousSharkTeeth061812.JPG#/media/File:CretaceousS
harkTeeth061812.JPG, 77. Megalodon tooth and great white shark teeth: "Megalodon tooth with great
white sharks teeth-3-2" by Megalodon_tooth_with_great_white_sharks_teeth-3.jpg:
*Megalodon_tooth_with_great_white_sharks_teeth.jpg: Brocken InagloryBlueRuler_36cm.png:
User:Kalanderivative work: Parziderivative work: Parzi - This file was derived from Megalodon tooth
with great white sharks teeth-3.jpg:. Licensed under CC BY-SA 3.0 via Commons -
https://commons.wikimedia.org/wiki/File:Megalodon_tooth_with_great_white_sharks_teeth-3-
2.jpg#/media/File:Megalodon_tooth_with_great_white_sharks_teeth-3-2.jpg, 78. Megalodon teeth:
Photo by Patrick Nurre, 78. Fossil megalodon teeth: "Megalodon teeth" by Catalina Pimiento, Dana J.
Ehret, Bruce J. MacFadden, Gordon Hubbell - Pimiento C, Ehret DJ, MacFadden BJ, Hubbell G (2010)
Ancient Nursery Area for the Extinct Giant Shark Megalodon from the Miocene of Panama. PLoS ONE
5(5): e10552. doi:10.1371/journal.pone.0010552.g002. Licensed under CC BY 2.5 via Commons -

https://commons.wikimedia.org/wiki/File:Dasyatis_brevicaudata_4x3.jpg#/media/File:Dasyatis_brevica
udata_4x3.jpg, 89. Eagle Ray: "Spotted Eagle Ray (Aetobatus narinari)2" by john norton - originally
posted to Flickr as Eagle Ray. Licensed under CC BY 2.0 via Commons -
https://commons.wikimedia.org/wiki/File:Spotted_Eagle_Ray_(Aetobatus_narinari)2.jpg#/media/File:S
potted_Eagle_Ray_(Aetobatus_narinari)2.jpg, 89.
Electric Ray: "Torpedo torpedo corsica2" by Roberto Pillon -
http://fishbase.org/photos/thumbnailssummary.php?ID=2062#. Licensed under CC BY 3.0 via
Commons -
https://commons.wikimedia.org/wiki/File:Torpedo_torpedo_corsica2.jpg#/media/File:Torpedo_torped
o_corsica2.jpg, 90. Sawfish: "Sawfish genova" by Flavio Ferrari - [1]. Licensed under CC BY-SA 2.0 via
Commons -
https://commons.wikimedia.org/wiki/File:Sawfish_genova.jpg#/media/File:Sawfish_genova.jpg, 90.
Ray fossil: From author's collection, photo by Patrick Nurre, 90. Fossil Ray tooth: From author's
collection, photo by Patrick Nurre, 91. Ray fossil barbed tails: From author's collection, photo by Patrick
Nurre, 91.

X.
Fossil turtle egg: From author's collection, photo by Patrick Nurre, 104. Diplodocus foot cast: Photo by
Vicki Nurre, 92. Sign: Photo by Vicki Nurre, 92. Disarticulated dinosaur bones: From author's
collection, photo by Patrick Nurre, 93. Articulated dinosaur: Photo by Patrick Nurre, 93. Hadrosaur:
Photo by Vicki Nurre, 94. Miscellaneous dinosaur bones: From author's collection, photos by Patrick
Nurre, 95. Dinosaur National Monument: Photo by Vicki Nurre, 96. Museum of the Rockies, t. rex:
Photo by Vicki Nurre, 96. Ceratopsian dinosaurs: Photos by Vicki Nurre, 97. Triceratops: "LA-
Triceratops mount-2" by Source: Allie_CaulfieldDerivative: User:MathKnight - File:LA-Triceratops
mount-1.jpg (by Allie_Caulfield). Licensed under CC BY-SA 3.0 via Commons -
https://commons.wikimedia.org/wiki/File:LA-Triceratops_mount-2.jpg#/media/File:LA-
Triceratops_mount-2.jpg, 98. Fossil frill bones and teeth: From author's collection, photos by Patrick
Nurre, 98-99. Hadrosaur eggs: Photos by Patrick and Vicki Nurre, 99. Dinosaur eggs: Photo by Patrick
Nurre, 99. Dinosaur egg shell shard: From author's collection, photo by Patrick Nurre, 100.
Disarticulated Dinosaur bones: From author's collection, photos by Patrick Nurre, 100. Disarticulated
dinosaur bones: From author's collection, photo by Patrick Nurre, 100. Cross-section of dinosaur bone:
From author's collection, photo by Patrick Nurre, 101. Gastroliths: From author's collection, photo by
Patrick Nurre, 101. Small dinosaur rib bones and digit bones: From author's collection, photos by Patrick
Nurre, 102. Dinosaur vertebrae bones: From author's collection, photos by Patrick Nurre, 102. Dinosaur
coprolite: From author's collection, photo by Vicki Nurre, 102. Herons and cranes: Photo by Vicki
Nurre, 103. Fossil turtle shell: Photo by Patrick Nurre, 103. Disarticulated fossil parts: From author's
collection, photo by Patrick Nurre, 103. Fossil turtle egg: From author's collection, photo by Patrick
Nurre, 104. Fossil crocodile scute: From author's collection, photo by Patrick Nurre, 104. Fossil alligator
tooth: From author's collection, photo by Patrick Nurre, 104. Fossil alligator vertebrae: From author's
collection, photo by Patrick Nurre, 105. Fossil snake vertebrae: From author's collection, photo by
Patrick Nurre, 105. Fossil mammal bone: From author's collection, photo by Patrick Nurre, 106. Other
mammal fossils: Photo by Patrick Nurre, 107. Fossil beaver tooth: From author's collection, photo by
Patrick Nurre, 107. Camel tooth: From author's collection, photo by Patrick Nurre, 107. Fossil
mammoth tusk: Photo by Patrick Nurre, 108. Mammoth tooth: Photo by Patrick Nurre, 108. Mammoth
skeleton: Photo by Vicki Nurre, 108. Fossils mammoth bone: From author's collection, photos by
Patrick Nurre, 108. Fossil tusk: From author's collection, photos by Patrick Nurre, 109. Disarticulated
mammoth tusk: From author's collection, photo by Vicki Nurre, 109. Mammoth vertebrae: From
author's collection, photo by Patrick Nurre, 109. Mastodon skeleton: Photo by Vicki Nurre, 110. Fossil
mastodon partial teeth: From author's collection, photos by Patrick Nurre, 110. Saber-toothed cat: Photo
by Vicki Nurre, 111. Fossil horse skeleton: Photo by Vicki Nurre, 111. Fossil horse tooth: From author's
collection, photo by Patrick Nurre, 112. Fossil deer antler and fossil mammal bone: From author's
collection, photo by Patrick Nurre, 112. Fossil small mammal vertebrae and fossil mammal bone: From
author's collection, photos by Patrick Nurre, 112. Fossil mammal coprolite: From author's collection,
photo by Patrick Nurre, 113. Large mammal vertebrae: From author's collection, photo by Patrick Nurre,
113.

XI.
Diatoms: Photo by Zach Kiser, used by permission, 114. Fossil cricket: From author's collection, photo
by Vicki Nurre, 114. Wyoming ant: From author's collection, photo by Patrick Nurre, 114. Ant in
limestone: From author's collection, photo by Patrick Nurre, 114. Wasp in limestone: From author's
collection, photo by Patrick Nurre, 114. Bee in limestone: From author's collection, photo by Patrick
Nurre, 115. Fly in amber: From author's collection, photo by Patrick Nurre, 115. Cricket in amber: From
author's collection, photo by Patrick Nurre, 115. Fossil larva in limestone: From author's collection,
photo by Patrick Nurre, 115. Insect swarm in limestone: From author's collection, photo by Patrick

Nurre, 115. Insect swarm in limestone: From author's collection, photo by Patrick Nurre, 116. Mosquito in amber: From author's collection, photo by Patrick Nurre, 116. Fossil wasp in limestone: From author's collection, photo by Patrick Nurre, 116. Fossil ferns and fossil reed: From author's collection, photos by Patrick Nurre, 117. Fossil pine cones: From author's collection, photos by Patrick Nurre, 117. Fossil palm branches: Photo by Vicki Nurre, 117. Fossil palm branch: Photo by Vicki Nurre, 118. Petrified wood: From author's collection, photo by Vicki Nurre, 118. Petrified wood: From author's collection, photo by Heidi Noggle, used by permission, 119. Petrified tree: Photo by Vicki Nurre, 119. Gravel pit: Photos by Vicki Nurre, 120. Fossil leaves: Photo by Vicki Nurre, 120. Mice fossils: Photo by Patrick Nurre, 121. Diatomaceous Earth: "Diatomaceous Earth BrightField" by Zephyris - Own work. Licensed under CC BY-SA 3.0 via Commons - https://commons.wikimedia.org/wiki/File:Diatomaceous_Earth_BrightField.jpg#/media/File:Diatomaceous_Earth_BrightField.jpg. Diatoms: Photo by Zach Kiser, used by permission, 122. Nummulitids: "Nummulitids" by Wilson44691 - Own work. Licensed under Public Domain via Commons - https://commons.wikimedia.org/wiki/File:Nummulitids.jpg#/media/File:Nummulitids.jpg, 122. Chert: From author's collection, photo by Heidi Noggle, used by permission, 122. Chalk: From author's collection, photo by Heidi Noggle, used by permission, 122. Flint: From author's collection, photo by Heidi Noggle, used by permission, 122. Euconodonta: "Euconodonta" by Philippe Janvier, 1997 - Tree of Life Web Project. Licensed under CC BY 3.0 via Commons - https://commons.wikimedia.org/wiki/File:Euconodonta.gif#/media/File:Euconodonta.gif, 123. Conodonts: "Conodonts". Licensed under Public Domain via Commons - https://commons.wikimedia.org/wiki/File:Conodonts.jpg#/media/File:Conodonts.jpg, 123.

Index of Fossil Pictures

manatee *(see dugong)*
marine reptile bones 82
mastodon 110
mastodon tooth 110
megalodon teeth 78-80
microfossil 121, 122
mold 29
mortality plate *(see also graveyard fossils)* 56, 57, 59, 61, 65, 72, 75
mosasaur 81
mosasaur teeth 81
mosquito in amber 116
mouse toe bone 121
mud crack 36
mussel 65
nautilus *(see ammonite)*
nummulitids *(fossil protozoa)* 75, 122
oyster 64
palm branch 117-118
pelecypod *(see clam, oyster)*
perch 86
petrified logs, wood 27, 34, 38, 40, 41, 118, 119
pine cones 117
protozoa *(nummulitid)* 75, 122
pseudofossil *(def. 32)* 33
puffer fish mouth part 87
ray 90, 91
ray fossil barbed tail 91
ray tooth 91
reed 117
ripple marks 37, 50
saber-toothed cat 111
sand dollar 59
scaphopod 21, 36,
sea urchin 58
shark teeth 77, 78, 79, 80
shark vertebrae 84
shrimp burrow 30

Patrick Nurre has been a rockhound since childhood. This early interest led him to a lifelong study of the world of geology. In 2005, he started Northwest Treasures, which is devoted to designing geology kits for schools, and he has written fourteen textbooks for the study of geology. He conducts numerous field trips each year in Washington State to such places as the Olympic Peninsula, Mt. Rainier, Mt. St. Helens, the Channeled Scablands, Mt. Baker and Whidbey Island. In addition, he also gives field trips to the volcano loop of Oregon and California, Mt. Hood volcanic area (Oregon), the eastern badlands of Montana and Yellowstone National Park. In 2012, he opened the Geology Learning Center in Mountlake Terrace, WA. Patrick is a popular speaker at homeschool conventions, schools, and churches. He currently co-pastors a church in the Seattle, Washington area. Patrick and his wife, Vicki, have three children and one grandchild, and live in Bothell, Washington.

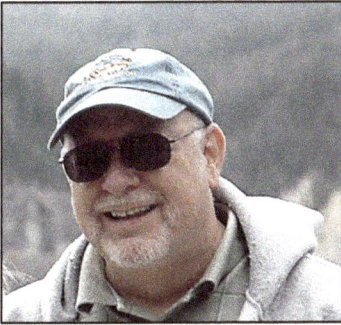

If you would like to contact Patrick about speaking or field trips: northwestexpedition@msn.com. For a list of speaking topics: NorthwestRockAndFossil.com

Other books by Patrick Nurre, available at
NorthwestRockAndFossil.com:

Bedrock Geology
Fossils and Dinosaurs for Little Eyes (PreK-3rd)
Fossils, Dinosaurs and Cave Men (high school)
Fossil Identification Made Easy (3-12 grades)
Genesis: Rock Solid
Geology for Kids (4-8 grades)
Geology and the Hawaiian Islands (5-12 grades)
Rock Identification Field Guide
Rock Identification Made Easy (3-12 grades)
Rocks and Minerals for Little Eyes (PreK-3rd)
Rocks and Minerals: The Stuff of the Earth (high school)
The Geology of Yellowstone: A Biblical Guide
Volcanoes for Little Eyes (PreK-3rd)
Volcanoes, Volcanic Rocks and Earthquakes (high school)

www.ingramcontent.com/pod-product-compliance
Lightning Source LLC
Chambersburg PA
CBHW042311210326
41598CB00041B/7347